CHEMICAL REACTOR DESIGN FOR PROCESS PLANTS

Volume Two

CHEMICAL REACTOR DESIGN FOR PROCESS PLANTS

Volume Two: Case Studies and Design Data

HOWARD F. RASE

W. A. Cunningham Professor of Chemical Engineering
The University of Texas at Austin

Original Illustrations by

JAMES R. HOLMES

Associate Professor of Engineering Graphics
The University of Texas at Austin

A WILEY-INTERSCIENCE PUBLICATION

JOHN WILEY & SONS, New York · London · Sydney · Toronto

Library of Congress Cataloging in Publication Data:

Rase, Howard F
  Chemical reactor design for process plants.

  "A Wiley-Interscience publication."
  Includes bibliographical references and index.
  CONTENTS: v. 2. Case studies and design data.
  1. Chemical reactors—Design and construction.
2. Chemical processes. I. Title.
TP157.R34   681'.766   77-1285
ISBN 0-471-01890-2 (v. 2.)

Printed in the United States of America

10 9 8 7 6 5 4 3

To my children

Carolyn Victoria

and

Howard Frederick, Jr.

PREFACE

This book has been written for the professional engineer who either daily or periodically must deal with design or operation of chemical reactors. But in addition to serving as a reference in the personal libraries of professionals, it should also be useful as a textbook for advanced design courses, including courses taught in continuing education programs.

The content, writing style, and arrangement are based on the needs of the competent professional engineer who not only seeks firm guidelines when confronted simultaneously with impending deadlines and complex design decisions but also desires full understanding of existing data and relevent theory. Only with such insight can he confidently make the difficult and demanding judgments essential for reliable design.

The development of the material, based on such a conceptual framework, logically fell into five parts:

In Part 1 principles of thermodynamics and reaction kinetics are discussed with special reference to data required for reactor design. Applicable theory, experimental techniques, and catalyst characteristics are major concerns that are presented in detail, along with many useful charts and tables, including such practical aids as guides for selecting commercially viable catalysts and detailed lists of catalyst poisons. Since the intellectually satisfying procedure of design involving kinetic models is not always possible or economical, techniques for gathering scale-up data are considered. Actual scale-up procedures are described in later sections for each reactor type.

General procedures for selecting reactor type and operating mode are considered in Part 2, along with certain design issues common to all reactors. Safety, reliability, and yield, together with the means for judging these important issues, are emphasized.

Major categories of reactors as applied to single-phase fluids are discussed in Part 3. Design procedures are given for homogeneous tubular, stirred (continuous and batch), fixed-bed, and fluidized-bed reactors. Since the chemical phenomena usually dominate, kinetic and thermodynamic data have been applied most successfully for these systems.

By contrast, reactions involving multiple-fluid phases, as described in Part 4, are often controlled by transport processes and therefore are more dependent on the exact character of the particular reactor type. Surface effects are important, and careful scale-up from pilot-plant data may be the only safe design procedure in many cases. Design for these multiphase systems is approached with particular reference to each unique reactor type, and scale-up procedures are strongly emphasized.

The ultimate test of all ideas, particularly in the technical world, is to use them in real problems. This final aspect is demonstrated by the 14 case studies contained in Part 5, Volume II. These studies all deal with industrially important reactions and reactor types and face directly the many difficult decisions that reality always imposes. In such an arena there is always room for debate, but in the interest of saving space, studies of alternatives have been limited, with the implied understanding that these can be handled similarly as the examples presented.

Volume I provides the essential principles and techniques, and Volume II presents examples of these as applied to design problems of industrial importance. The point of division into two volumes was selected as a means of enhancing the usefulness of each. The case studies, which are referred to frequently in Volume I, and the nomenclature and appendixes were placed in Volume II. This arrangement is convenient for side-by-side use of both volumes, since it avoids the necessity of flipping from text material to the back of Volume I to verify nomenclature or examine a relevant case study or appendix. However, the nomenclature is also included in Volume I to make it self-contained.

Although the material has been arranged in a logical order as just described, great care has been exercised to make it possible for the user to enter the volumes at any point required by the needs and interests of the moment. In so pursuing a particular topic of interest, one will be led by cross-references to discussions and theoretical developments in related portions of the book and also to the wealth of other literature by numerous citations of journals and important books.

I am indebted to a number of former graduate students at The University of Texas who as Teaching Assistants in my design courses aided in developing computer programs for student use, some of which were adapted for the Case Studies.

I appreciate most sincerely the advice of my colleagues, Professors M. Van Winkle and D. R. Paul, and my former colleague, Dr. G. W. Dowling, who originated several of the more complex computer programs. Dr. J. R. Fair of Monsanto has encouraged me from the beginning in my desire to serve the needs of the professional.

HOWARD F. RASE

Austin, Texas
January 1977

CONTENTS

Volume Two

---

* All nomenclature is defined in this section. It is also defined in the text at point of first use.

CHEMICAL REACTOR DESIGN
FOR PROCESS PLANTS

Volume Two

INTRODUCTION

to Volume Two

THIS VOLUME contains 14 case studies that illustrate many of the principles and techniques discussed in Volume I. The Appendix material, referenced in both volumes, is included to provide frequently used data for design purposes. The reader will often find it convenient to use Volume I and Volume II side by side. Since frequent references are made to the case studies in Volume I, simultaneous study of both the basic concepts and the applications will prove beneficial. The complete nomenclature list is given in each volume so that the reader may refer to the volume not in direct use at any particular time rather than flipping to the back of the one being studied.

PART FIVE

Case Studies

ASSEMBLED IN THIS SECTION are 14 case studies designed to illustrate some of the major principles discussed in Volume I, Parts I–IV. When dealing with real problems, one is faced with numerous compromises, lack of data, and decisions relative to economics, ecology, and safety. Some of the problem statements have been couched in a form that does not preselect the reactor type, so that this decision process may be illustrated along with a logical approach to kinetics and chemistry of the system. Economic studies have been illustrated in several cases, but in these days of rapidly changing costs only the techniques illustrated may be viewed with any equanimity. Mechanical designs are not pursued although process vessel sketches are developed in several cases. Only one pure scale-up problem is demonstrated, which is in no way proportional to the utility of this technique.

The weakest aspect of these designs is the data on which they are based including both kinetic data and certain physical properties such as viscosity, a crucial factor in some heat-transfer calculations. Only data in the open literature have been used and no assurance of the validity of the final calculations for specific systems can be given. The techniques here demonstrated, however, when applied to good data for specific systems should yield valid results for reasonable design decision.

Brief descriptions of algorithms for computer calculations have been given, except when standard differential-equation solving routines are used. Those algorithms that are described are not necessarily the most efficient, and one is referred to standard works on computer programming for such important issues.

Many engineers must continue to work in mixed systems of units, and the case studies reflect this by employing both British and metric units depending on the original data and often customs (U.S.A.) of the particular industry related to the case study.

$$C=C + \bigcirc \longrightarrow \overset{C-C}{\bigcirc} \underset{-H_2}{\longrightarrow} \overset{C=C}{\bigcirc}$$

CASE STUDY 101

Styrene Polymerization

THIS STUDY is a nontrivial case of a CSTR involving the complexities of free-radical polymerization. The weakest aspect of the model is the method for estimating viscosity, which strongly affects heat transfer and is very sensitive to temperature and concentration of polymer. Experimental work with the particular polymer-styrene product mix would yield a better correlation.

Problem Statement

Design the first stage of a two-stage thermal polymerization system for producing 40 million lb/yr of polystyrene with an overall conversion for the system of 95%. The product from the first stage should have a number average molecular weight $(\overline{M}_n)$ of 144,000.

Feed. 99.5 wt% styrene with 10 ppm maximum polymer and maximum weight percentages of impurities of 0.02 aldehydes as CHO, 0.01 peroxide as H_2O_2, 0.0025 sulfur, and 0.01 chloride.

Chemistry and Kinetics

Details of the formation of a Diels–Alder adduct (AH) and its radical (A ·) from styrene monomer (M) have been described (1) and are partially illustrated as follows, together with the propagation, termination, and chain-transfer steps.

5

Initiation

$$M + M \underset{k_{-1}}{\overset{k_1}{\rightleftharpoons}} \quad \text{(call AH)} \tag{1}$$

$$M + AH \xrightarrow{\ k_2\ } M\cdot + A\cdot \tag{2}$$

$$M + AH \xrightarrow{\ k_3\ } \text{trimer} \tag{3}$$

$$A\cdot + M \xrightarrow{\ k_4\ } R_1\cdot \tag{4}$$

$$M\cdot + M \xrightarrow{\ k_5\ } R_1\cdot \tag{5}$$

Propagation

$$R_r\cdot + M \xrightarrow{\ k_{pr}\ } R_{r+1}\cdot \tag{6}$$

Termination by Combination

$$R_r\cdot + R_s\cdot \xrightarrow{\ k_t\ } P_{r+s} \tag{7}$$

Chain Transfer.   (to AH or M)

$$R_r\cdot + AH \xrightarrow{\ k_{tr}\ } P_r + R_1\cdot \tag{8}*$$

The rate of initiation is

$$r_1 = (k_4[A\cdot] + k_5[M\cdot])[M] \tag{CS-1.1}$$

By using the stationary state hypothesis, expressions for concentrations of $A\cdot$, $M\cdot$, AH, and $R_r\cdot$ can be obtained as functions of monomer concentrations so that Eq. CS-1.1 can also be written in these terms (1)

$$(r_1) = \frac{2k_1k_2[M]^3}{k_{-1} + (k_2 + k_3)[M] + k_{tr}(r_1/k_t)^{\frac{1}{2}}} \tag{CS-1.2}$$

where [M] is the concentration of monomer in g moles/liter. A limiting case in which k_{-1} is much greater than the remainder of the denominator appears to fit operating data.

$$(r_1) = \frac{2k_1k_2}{k_{-1}}[M]^3 = 2k_i[M]^3 \tag{CS-1.3}$$

---

* AH represents a Diels–Alder adduct.

Other instantaneous expressions typical of free-radical polymerization with termination by combination dominating are (1).

$$r_{pr} = k_{pr}[R_r \cdot][M] = k_{pr}\left(\frac{r_1}{k_t}\right)^{\frac{1}{2}}[M] \qquad \text{(CS-1.4)}$$

From Eq. CS-1.3

$$r_{pr} = \left(\frac{2k_i}{k_t/k_{pr}^2}\right)^{\frac{1}{2}}[M]^{\frac{3}{2}} = A[M]^{\frac{3}{2}} \qquad \text{(CS-1.5)}$$

The grouping A has been found to vary with conversion.

$$A = A_0 \exp(A_1 X + A_2 X^2 + A_3 X^3) \qquad \text{(CS-1.6)}$$

Also

$$(\overline{DP})^{-1} = C_m + \frac{r_1}{2r_{pr}} \qquad \text{(CS-1.7)}$$

or

$$(\overline{DP})^{-1} = C_m + \frac{k_i[M]^{\frac{1}{2}}}{A} \qquad \text{(CS-1.8)}$$

$$\overline{M}_n = 104\,\overline{DP} \qquad \text{(CS-1.9)}$$

where $\overline{M}_n$ is the number average molecular weight. C_m, which accounts for chain transfer, also varies with conversion.

$$C_m = C_{m_0} + B_1 X \qquad \text{(CS-1.10)}$$

Equations for the several constants are given in Table CS-1.1.

Table CS-1.1 Kinetic Data for Thermal Polymerization of Styrene (Ref. 1)

$A_0 = 1.964 \times 10^5 \exp(-10{,}040/T), (1./\text{g mole})^{\frac{3}{2}} \sec^{-1}$

$A_1 = 2.57 - 5.05 \times 10^{-3}T$

$A_2 = 9.56 - 1.76 \times 10^{-2}T$

$A_3 = -3.03 + 7.85 \times 10^{-3}T$

$B_1 = -1.013 \times 10^{-3} \log_{10}\left(\frac{473.12 - T}{202.5}\right)$

$k_i = 2.19 \times 10^5 \exp(-13{,}810/T), (1./\text{g-mole})^2 \sec^{-1}$

$C_{m_0} = 2.198 \times 10^{-1} \exp(-2820/T)$

$T = {}^\circ K$

Thermodynamics

The typical high exothermicity of polymerization, $\Delta H = -17{,}500$ cal/g mole monomer converted @ 25°C for this case (2,3), and the importance of temperature in controlling molecular weight indicates a definite need for a reactor with heat transfer. From Table 6.1, pp. 270–271[1], the adiabatic factor is moderately high, and the heat generation potential is not high.

Reactor Type

A CSTR has good heat-transfer characteristics and allows continuous operation, but it can only be used to a conversion level where heat transfer and contacting become poor because of increasing viscosity. This problem can be overcome by diluting with solvent, but then solvent handling and recovery add greatly to the cost. Alternatively, a CSTR can be specified for some low conversion range, and the remaining conversion (up to 95) can be completed in a screw-type, tubular flow unit, such as described on p. 480[1], or as shown in Fig. 10.29. We will only consider the CSTR, for the tube operates at such high viscosity that diffusion controls and modeling techniques are inadequate. Styrene polymerizes slowly relative to many other monomers and a CSTR is also advantageous for this reason.

Design Model (CSTR)

Equations CS-1.5–CS-1.10 express instantaneous values applicable to a CSTR. The following mole and heat balances apply.

Monomer Balance.   (neglecting monomer used in initiation and transfer)

$$F_M X_M = \frac{-d[M]}{dt} V = r_{pr} V = A[M]^{\frac{3}{2}} V \qquad \text{(CS-1.11)}$$

Heat Balance

$$104 F_M c_{pM}(T_e - T_0) + U A_h(T_e - T_j) = (-\Delta H_M)_{T_e}(F_M X_M) \qquad \text{(CS-1.12)}$$

where F_M is the molar flow rate of monomer fed, c_{pM} is the heat capacity of monomer, and T_e, T_0, T_j are the exit, entrance, and cooling medium temperatures, respectively.

Since the heat capacities of monomer and polymer are approximately the same, the heat of reaction at 25°C can be used in Eq. CS-1.12.

Fluid Properties

$$c_{p_M} = 0.353 + (0.0014)(T - 293), \text{cal/g}°\text{C} \qquad \text{(CS-1.13) (2,3)}$$

$$\rho_M = 924 - 0.918(T - 273.1), \text{g/l.} \qquad \text{(CS-1.14) (1)}$$

$$\rho_{pr} = 1084.8 - 0.605(T - 273.1), \text{g/l.} \qquad \text{(CS-1.15) (1)}$$

$$\delta_v = \left(\frac{1}{\rho_{pr}} - \frac{1}{\rho_M} \right) \bigg/ \frac{1}{\rho_M} \qquad \text{(CS-1.16)}$$

$$[M] = \frac{\rho_M(1 - X_M)}{(104)(1 + \delta_v X_M)}, \qquad \text{g moles/l.} \qquad \text{(CS-1.17)}$$

$$(\eta_r - 1)/w_{pr} = 109[\eta]_T \exp\{2079 w_{pr}(1.09[\eta]_T + 1)/T\} \quad \text{(CS-1.18) (4)}$$

where ρ_M and ρ_{pr} are the densities of monomer and polymer, respectively. η_r is the ratio of viscosity of polystyrene in styrene to that of pure styrene at the same temperature, $[\eta]_T$ is the intrinsic viscosity of polystyrene of same molecular weight and type in toluene at 30°C in dl./g, and w_{pr} is the weight fraction as mass of polymer/mass of styrene.

The viscosity is very sensitive to temperature and concentration. Equation CS-1.18 is corrected to agree with the original plot of experimental data given in the reference. Ideally the value of $[\eta]_T$ should be determined for the product in question in the laboratory. For illustrative purposes we select a value of 0.7.

Design Calculations and Decisions

Operating Temperature and Conversion

Molecular weight in thermal polymerization is most strongly affected by operating temperature and is not sensitive to conversion. By applying Eqs. CS-1.6–CS-1.9 at various temperatures using a hand calculator, a temperature of 150°C gave $\overline{M}_n = 144{,}041$ @ $X = 0.4$ and $\overline{M}_n = 143{,}621$ @ $X = 0.45$. The corresponding solution viscosities at these two conversions are 3623 cp and 13406 cp, respectively. Styrene viscosity at 150°C is 0.22 cp (5). Thus it is seen that viscosity increases dramatically above $X_M = 0.4$, and heat-transfer rate will decline rapidly such that it will not be possible to remove the heat of polymerization. Hence for safety reasons the design will be set at $X_M = 0.4$ as a maximum, and subsequent calculation will reveal that this is operable.

Monomer Flow Rate

Basis: 95% overall conversion for 2-stages.
     95% operating factor
     99.5% styrene purity and no recycle

$$\frac{4 \times 10^7 \text{ lb/yr}}{(0.995)(0.95)(0.95)(365)(24)(3600)} = 1.413 \text{ lb/sec or } 641 \text{ g/sec}$$

Reactor Volume

From Eq. CS-1.11

$$V = \frac{F_M X_M}{A[M]^{\frac{3}{2}}} = \frac{(641/104)(0.4)}{(1.64243 \times 10^{-5})(4.9506)^{\frac{3}{2}}}$$

$$= 2752.7 \text{ liters or } 727 \text{ gal}$$

A 750 gal stainless-steel tank (60 × 60 in.) is selected from Table 8.8, $D = 60$ in., $D_1 = (1/3)60 = 20$ in. with 160 rpm for good heat transfer (838 ft/min tip speed, see Table 8.10). One 20-in. flat-blade turbine is selected initially for analysis.

Heat Balance (Eq. CS-1.12)

Basis: 150°C which is the reaction temperature
     30°C inlet, 45°C average cooling water temperature.
     $(641)(0.451)(150 - 30) + UA_h(150 - 45) = (17,500)(641/104)(0.4)$

Required $UA_h = (43144 - 34691)/105 = 80.5$ cal/sec °C,

$$U = \frac{(80.5)(3600)}{(97)(727/750)(252)(1.8)} = 6.8 \text{ BTU/hr ft}^2$$

Estimate h and U. From Eq. 8.14

$$h = \frac{0.73\lambda_f}{D}\left(\frac{c_p \mu}{\lambda_f}\right)^{0.33}\left(\frac{\rho N D_1^2}{\mu}\right)^{0.65}\left(\frac{\mu}{\mu_w}\right)^{0.24}$$

$$= \frac{0.73}{20/12}(0.055)\left[\frac{(0.451)(3623)(2.42)}{0.055}\right]^{0.33}$$

$$\times \left[\frac{(53.5)(160)(60)(20/12)^2}{(3623)(2.42)}\right]^{0.65}\left(\frac{\mu}{\mu_w}\right)^{0.24} = 26.4\left(\frac{\mu}{\mu_w}\right)^{0.24}$$

$$\bar{\rho} = \frac{\rho_M/1000}{1 + \delta_r X_M} = \frac{0.786}{1 - (0.209)(0.4)} = 0.858 \text{ g/cm}^2 \text{ or } 53.5 \text{ lb/ft}^3$$

$\lambda_f = 0.055$ BTU/(hr)ft^2°F ft^{-1} @ 150°C for ethylbenzene (*API Data Book*)

Estimation of wall temperature ($b_w = \frac{5}{16}$ in., fouling factor, $\mathscr{R}_h = 0.001$, h_0 is for water $= 400$ BTU/hr ft^2 °F.

$$h(T - T_w) = \frac{1}{(1/h_o) + (b_w/\lambda_w) + \mathscr{R}_h} (T_w - T_j)$$

$$\frac{T_w - T_j}{T - T_w} = h\left(\frac{1}{h_o} + \frac{b_w}{\lambda_w} + \mathscr{R}_h\right) = h\left(\frac{1}{400} + \frac{0.026}{30} + 0.001\right)$$

$$= 4.367 \times 10^{-3}h$$

Assuming $h = 10$

$$\frac{T_w - 45}{150 - T_w} = 0.044 \qquad \text{or} \qquad T_w = 49°C$$

for which μ from Eq. CS-1.18 is 52,477 cp

$$\text{corrected } h = 26.4\left(\frac{3623}{52,477}\right)^{0.24} = 13.9$$

$$U = \frac{1}{9.98} + 4.367 \times 10^{-3} = 13.1 \text{ BTU/hr ft}^2 \text{ °F}$$

Thus the design is adequate compared with required U of 6.8. It should be noted that an increase in conversion to 45% at a lower flow rate to yield the same production would produce a h equal to 6.95 or a U of 6.75, based on the viscosity correlation of Eq. CS-1.18. The required U at this new condition would be 9.9. Thus somewhere between 40% and 45% conversion the reaction will become uncontrollable by conventional jacket cooling, but the reaction is slow and corrections can be made. The vapor pressure of styrene at 150°C is 880 mm Hg (2), but at the conversion level the actual value will be less because of nonidealities attributable to the dissolved polymer. An approximate operating pressure of 2.5 psi will be assumed. This pressure or higher values at higher temperatures which could occur during upsets can be released to cause vaporizing of the monomer and rapid cooling.

Required Horsepower

Based on manufacturer's correlation Fig. 8.8 at $N_{Re} = 163$, $N_p = 5.5$

$$\text{hp} = (1.1)(3.52 \times 10^{-3})(5.5)\left(\frac{53.5}{62.4}\right)\left(\frac{160}{60}\right)^3\left(\frac{20}{12}\right)^5 + 0.5 = 4.95 \text{ hp or}$$

specify 5 hp motor.

If one refers to Fig. 8.7, the Reynolds number is in the transition region for which the power number is 3.5 and the corresponding hp $= 3.3$; but the higher hp would be selected since the reactor must be periodically washed with a low viscosity solvent which would place the operation in the turbulent region with a higher power number.

Mixing Effectiveness

Equation 8.16 can be used to estimate the effective radius of agitation.

$$R_{eff} = c\sqrt{\frac{hp}{\mu}} = c\sqrt{\frac{4.05}{3623}} = 0.033c$$

Using appropriate values of the coefficient, horizontal and vertical values are 4.2 and 1.7 ft, respectively, which is adequate in the horizontal but not the vertical direction for a 60 × 60 in. tank. Accordingly, two 18 in. impellers should be substituted. By selecting the 18 in. impellers it is possible to maintain the required power at the same value, $(18/20)^5(0.9)(2) \approx 1.0$ (see instructions in caption for Fig. 8.7, p. 348[1]). The turbines should be spaced 30 in. apart.

A better decision would be to use a single curved-blade turbine (with six blades) since it provides superior axial mixing. The resulting hp will be lower, but the heat-transfer coefficient will be approximately the same (see Table 8.6). Because of the flatter power-number curve (Fig. 8.7), the power overage required for washing will not be as great. Manufacturer's representatives should be consulted for the final selection.

REFERENCES

1. A. W. Hui and A. E. Hamielec, *J. Appl. Polym. Sci.*, **16**, 749 (1972).
2. K. E. Coulter, H. Kehde, and B. F. Hiscock, in *High Polymers*, Vol. 24, Part 2, E. C. Leonard (ed.), Wiley-Interscience, New York, 1971, p. 501.
3. R. H. Boundy and R. F. Boyer, *Styrene*, ACS Monograph, Reinhold, New York, 1952.
4. M. Hirose, E. O'Shima, and H. Inoue, *J. Appl. Polym. Sci.*, **12**, 9 (1968).
5. K. E. Coulter, H. Kehde, and B. F. Hiscock, in *High Polymers*, Vol. 24, Part. 2, E. L. Leonard (ed), Wiley-Interscience, New York, 1971, p. 499.

CASE STUDY 102

Cracking of Ethane to Produce Ethylene

THIS STUDY, divided into three parts, illustrates the three levels of design models described on p. 456[1]. It also demonstrates design techniques for direct-fired tubular reactors.

With the rapid increase in raw-material costs and shortages of light hydrocarbons, feeds such as naphtha, gas oils, and even crude oil are being used for olefin manufacture. Ethane is an ideal feed if ethylene alone is the desired product. Other olefins, aromatics, and various additional products result when heavier hydrocarbons are cracked. The techniques of design demonstrated here are applicable to other light hydrocarbons, and the method of Case 102B is readily adaptable to heavier hydrocarbon systems such as naphtha and gas oil.

The weakest aspect in all the design models illustrated is the lack of an accurate quantitative means for predicting coking rate.

Problem Statement

Design a reactor system for producing 2.7 million lb/stream day of polymer grade ethylene using commercial ethane as feedstock. Ethane arrives at the battery limits in the liquid state at 700 psig and 80°F.

Feed to Unit		Final Product Specifications	
Component	Mole %	Component	Mole %
CH_4	3	CH_4	0.6
C_2H_6	94	C_2H_6	0.4
C_3H_8	3	C_2H_4	99.0
CO_2	800 ppm	C_2H_2	<10 ppm
S	15 grains/100 scf	H_2S	<5 ppm
		CO_2	<100 ppm

Chemistry

The pyrolysis of ethane occurs primarily by a homogeneous, free-radical chain reaction although important heterogeneous wall effects exist. It is now possible to predict homogeneous mechanisms for any higher paraffin from existing data on C_2–C_5 paraffins and to extend successfully data derived in batch at low temperatures to the high temperatures of industrial interest (1). For consistency such data must be taken in the absence of oxygen and carbon deposits on the walls (2,3). Both can accelerate the reaction, as can metal oxide coatings occurring on steel tubes (3). These surface effects, of course, can be most pronounced on laboratory equipment, and must be eliminated to obtain true homogeneous kinetics.

The generally accepted mechanism for ethane pyrolysis above 650°C is as shown in Fig. CS-2.1(1). Below 640°C secondary propylene formation occurs by another mechanism (4). The mechanistic scheme shown in Fig. CS-2.1 has been confirmed with laboratory data obtained at conversions less than 20% to avoid coke deposition on the reactor surface (1).

Extending this reaction scheme to a real system involving higher conversions and wall effects becomes a less certain exercise, and inevitably involves empiricism. Since in only the past decade has the homogeneous pyrolysis clearly been defined, it is understandable that resort to totally empirical rate forms has been and still is often the preferred method especially for complex feeds such as naphtha.

			A	E, kcal/g mole
Initiation	$C_2H_6 \rightarrow 2CH_3\cdot$	(1)	1.0×10^{16}	86
Propagation	$\begin{cases} CH_3\cdot + C_2H_6 \rightarrow CH_4 + C_2H_5\cdot \\ C_2H_5\cdot \rightarrow C_2H_4 + H \\ H\cdot + C_2H_6 \rightarrow H_2 + C_2H_5\cdot \end{cases}$	(2) (3) (4)	3.16×10^8 3.98×10^{13} 1.25×10^{11}	10.8 38 9.7
Termination	$\begin{cases} C_2H_5\cdot + C_2H_5\cdot \rightarrow n\text{-}C_4H_{10} \\ C_2H_5\cdot + C_2H_5\cdot \rightarrow C_2H_6 + C_2H_4 \end{cases}$	(5a) (5b)	2.511×10^{10}	0
Propylene formation	$C_2H_5\cdot + C_2H_4 \rightarrow C_3H_6 + CH_3\cdot$	(6)	3.16×10^9	19
	$\begin{bmatrix} C_2H_5\cdot + C_2H_4 \rightleftarrows 1\text{-}C_4H_9\cdot \\ 1\text{-}C_4H_9\cdot \rightarrow 2\text{-}C_4H_9\cdot \\ 2\text{-}C_4H_9\cdot \rightarrow C_3H_6 + CH_3\cdot \end{bmatrix}$			
Inhibition	$H\cdot + C_2H_4 \rightarrow C_2H_5\cdot$	(7)	5.0118×10^{10}	-0.8

Units for $A = sec^{-1}$ for Reactions 1 and 3, and $l.(g\ mole)^{-1}sec^{-1}$ for others.

Fig. CS-2.1 Free-radical mechanism for ethane cracking. [Reaction 5a can be neglected in terms of product produced and a combined rate constant used, $k_5 = 1.15k_{5a}(1)$. Rate equations in terms of partial pressures may be written using $R'T$ corrections.]

The chain initiating step, reaction 1, and the primary product forming step, reaction 3, are close to first order and have high activation energies. The other major steps shown are second order and have lower activation energies. The reactions yielding higher hydrocarbons and coke involve combinations and polymerizations, all of which would be reactions of orders higher than one. Experimental and plant data indicate that these reactions also have lower activation energies.

Conclusions (applicable to light hydrocarbon cracking):

1. A rapidly rising temperature profile will favor the primary decomposition to ethylene over other steps leading to higher hydrocarbons and coke.

2. Low reactant partial pressure favors primary reactions leading to ethylene over secondary unwanted reactions. Thus the use of an inert gas such as steam is indicated.

3. Short reaction contact time and plug flow will minimize unwanted secondary reaction.

Thermodynamics

Simultaneous equilibrium calculations for the several chain reactions are not productive, but one may usefully consider the apparent equilibrium between ethane, ethylene and hydrogen. In fact, such apparent equilibria are established at adequate contact time for many types of pure and mixed hydrocarbon feeds (5). Standard free energies of formation as plotted in Fig. 1.9 for ethane and ethylene indicate that a temperature in the range of $1100°K$ ($\approx 1500°F$) will be essential for a reasonable amount of conversion. In this range the apparent heat of reaction is highly endothermic ($34.56 \, kcal/g$ mole). The high endothermicity and high activation energy are reflected in the large negative adiabatic factor (-979) and negative heat generation potential (-29.4) as shown in Table 6.4.

Conclusions:

1. The reaction will demand large quantities of heat at high-temperature levels indicating the following reactor types.
   (a) Direct-fired furnace
   (b) Pebble heater
   (c) Fluidized bed
   (d) Direct-contact molten lead
   (e) Autothermic cracking
   (f) Regenerative furnace

The advent of tubing for direct-fired furnaces which can withstand the high temperatures required for steam cracking has essentially eliminated types (b) through (f). These were all developed to avoid heat transfer through metal walls. As heavier oils are cracked, however, fluidized beds may prove attractive because of the ease of transforming coke to a useful fuel gas.

2. Since the reaction produces an increase in moles, low pressures and the presence of inerts will favor higher equilibrium conversions. Steam is particularly attractive as an "inert," for it also tends to scavenge coke deposits or suppress coke formation by the water–gas reaction ($C + H_2O \rightleftarrows CO + H_2$).

Kinetics

This case study is useful beyond the limits of the particular reaction type, for it is helpful in illustrating the application of three types of data to a complex reaction system such as is typified by a free-radical mechanism. These three approaches are designated as Case 102A, 102B, and 102C.

Case 102A

As discussed on p. 456[1], the rate of disappearance of the primary reactant in light hydrocarbon pyrolysis may be expressed as a first-order rate equation. The product distribution as a function of conversion of key reactant is obtained from pilot-plant or full-scale plant data (6), and plotted as shown in Fig. 10.15, p. 448[1]. These data have been fitted to polynomials, as shown in Table CS-2.1. Since commercial operation occurs over a relatively narrow

Table CS-2.1 Empirical Product Distribution Equations

$n_2 = 16.7\, m_1 + 35\, m_2 + 52\, m_3 + 64\, m_4$

$n_3 = 1.0\, m_1 + 3.7\, m_2 + 6\, m_3 + 9\, m_4$

$n_4 = 2.143 X_A$

$n_5 = 1.429 X_A$

$n_6 = 16\, m_1 + 37\, m_2 + 56\, m_3 + 75\, m_4$

$n_7 = $ (total mass of ethane fed minus mass of components 1–6)/(MW of C_5's)[a]

where n = moles of indicated component per 100 moles of ethane fed,

X_A = conversion of ethane, moles ethane conv/mole fed, component, 2 = ethylene, 3 = methane, 4 = C_3's, 5 = C_4's, 6 = H_2, and 7 = C_5's.

$m_1 = -104.17 X_A (X_A - 0.4)(X_A - 0.6)(X_A - 0.8)$

$m_2 = 156.25 X_A (X_A - 0.2)(X_A - 0.6)(X_A - 0.8)$

$m_3 = -104.17 X_A (X_A - 0.2)(X_A - 0.4)(X_A - 0.8)$

$m_4 = 26.04 X_A (X_A - 0.2)(X_A - 0.4)(X_A - 0.6)$

Based on Fig. CS-2.2, low outlet pressure range (8–10 psig) and tabular data in Ref. 6.
[a] Taken as 71.1

temperature and pressure range, the product distribution is assumed independent of pressure and temperature over narrow ranges. This assumption is the weakest element of the technique, and the design conditions must be in the same range of pressure, temperature, and conversion as the experimental ractor. Thus the method does not allow much freedom in investigating design variables. It does, however, enable one to predict temperature, pressure, and conversion profiles for various sizes of reactors thereby providing a rational scale-up procedure (12).

The first-order rate constant for ethane is obtained from Table 10.4, p. 446[1].

$$k_c = 4.717 \times 10^{14} \, e^{-72,240/R'T}, \, \text{sec}^{-1} \qquad \text{(CS-2.1)}$$

where $T = {}^\circ$K or

$$k_p = \frac{k_c}{RT} = \frac{4.717 \times 10^{14}}{RT} e^{-130,032/R'T}, \frac{\text{lb mole}}{(\text{atm})(\text{ft}^3)(\text{sec})} \qquad \text{(CS-2.2)}$$

where $T = {}^\circ$R and $R = 0.73$.

Case 102B

$$k_c = A \, e^{-E/RT}$$

$A \, [=] \, \text{sec}^{-1}$

$E \, [=] \, \text{kcal}/\text{g mol}$

The concept of apparent overall reactions as an empirical substitute for a free-radical mechanism was introduced in the 1940s (7). Stoichiometric expressions for the reactions are selected to produce the primary products observed experimentally, and rate equations are obtained by trial-and-error so that when used in a stepwise integration, the original experimental results are reproduced. Background on procedures for interpreting experimental data for complex reactions are given in Chapter 5.

Before the days of computers, the hand calculations were so laborious both for obtaining the data and then applying it to design that the procedure received little attention. With the advent of computers, however, the technique has become the most widely employed procedure for designing reactors of this type. If sufficient data were obtained experimentally, it is possible to calculate the effect of varying design conditions on product distribution and reactor size. Such freedom for calculating allows one to seek an optimum design basis.

For ethane pyrolysis several schemes of apparent reactions and rate expressions are reported (8,9). The scheme shown in Table CS-2.2 was selected for this case study (8): similar data have been reported for propane pyrolysis (10), but the rate constants have not been completely defined.

Although reaction 6 produces carbon, this does not account for all the coke that is formed. No satisfactory means for accurately calculating coke buildup has been reported, although proprietary empirical equations are used by companies having a great deal of operating data on a particular type of

Table CS-2.2   Apparent Reaction Scheme and Kinetics—Alternate B (8)

1. $C_2H_6 \rightleftarrows C_2H_4 + H_2$

$$r_1 = \frac{k_{p_1}P}{\mathscr{F}_T}\left[\mathscr{F}_{C_2H_6} - \frac{\mathscr{F}_{C_2H_4}\mathscr{F}_{H_2}P}{K_{p_1}\mathscr{F}_T}\right]$$

2. $C_2H_4 + 2H_2 \rightleftarrows 2CH_4$

$$r_2 = \frac{k_{p_2}P}{\mathscr{F}_T}\left[\frac{P}{\mathscr{F}_T}\mathscr{F}_{C_2H_4}\sqrt{\mathscr{F}_{C_2H_6}\mathscr{F}_{H_2}} - \frac{\mathscr{F}_{CH_4}}{K_{p_2}}\right]$$

3. $C_2H_4 \rightarrow 0.25C_4H_6 + 0.125C_4H_8 + 0.125C_4H_{10} + 0.125H_2$ $r_3 = 0.012r_1P$

4. $C_2H_4 \rightarrow 0.333C_6H_6 + H_2$

$$r_4 = \frac{k_{p_4}P^2}{(\mathscr{F}_T)^2}(\mathscr{F}_{C_2H_4})^2$$

5. $C_2H_4 \rightarrow C_2H_2 + H_2$ $r_5 = \dfrac{k_{p_5}P}{(\mathscr{F}_T)^2}(\mathscr{F}_{C_2H_4})^2$, (first order with respect to pressure)

6. $C_2H_4 \rightarrow 2C + 2H_2$

$$r_6 = \frac{k_{p_6}P^2}{(\mathscr{F}_T)^2}(\mathscr{F}_{C_2H_4})^2$$

7. $C_2H_4 + C_2H_6 \rightarrow 0.952C_3H_6 + 0.381C_3H_8 + 0.62H_2$

$$r_7 = \frac{k_{p_7}P}{\mathscr{F}_T}\left[\mathscr{F}_{C_2H_6} - \frac{\mathscr{F}_{C_2H_4}\mathscr{F}_{H_2}P}{K_{p_1}\mathscr{F}_T}\right]$$

Rate constants ($T = {}^\circ$K):

$$k_{p_1} = (3.59 \times 10^{14}/T)\,e^{-36356/T}{}^a$$
$$k_{p_2} = 7.50 \times 10^7\,e^{-29,300/T}$$
$$k_{p_4} = 4.09 \times 10^9\,e^{-33,400/T}$$
$$k_{p_5} = 9.13 \times 10^7\,e^{-30,800/T}$$
$$k_{p_6} = 4.51 \times 10^4\,e^{-24,500/T}$$
$$k_{p_7} = 1.04 \times 10^{14}\,e^{-46,200/T}$$

Units are lb moles/ft^3 sec atm for k_{p_1} and k_{p_5}, and lb moles/ft^3 sec(atm)^2 for all others.
Equilibrium constants ($T = {}^\circ$K):

$$K_{p_1} = 3.31 \times 10^{-7}\,e^{0.014T}$$
$$K_{p_2} = 1.0$$

[a] Corrected in accordance with Fig. 2 in Ref. 8.

furnace. In lieu of such data it has been suggested that conditions be avoided known to produce coke rapidly (8). The equilibrium approach criteria, discussed on p. 441[1], may be used for this purpose to provide a conservative safety margin. If a rate equation for total coke formation is known, the designer can calculate local rates of coke buildup, just as he is able to calculate rates for other products using empirical rate equations similar to those previously presented. This possibility frees him to consider more severe conditions in some portions of the coil that would not be permitted by the equilibrium-approach criteria. It would seem, however, that the most accurate rates of coke formation would require a complete free-radical mechanism that would include all major coke-producing steps.

Case 102C

The free-radical scheme shown in Fig. CS-2.1 is used for this alternate. Because it is closest to the actual phenomena, theoretically one could study operating variables over the broadest range. Major uncertainties exist, however, because of inadequate understanding of wall effects particularly related to coke formation.

General Design Conditions and Data for Alternates

Feed Composition and Flows

Basis: 2.7 million lb/day of 99 mole % purity

Product	Mole Fraction	MW	$(y)(MW)$
CH_4	0.006	16.04	0.096
C_2H_6	0.004	30.07	0.120
C_2H_4	0.99	28.05	27.770
	1.000	$MW =$	27.986

$$\frac{(2.7 \times 10^6 \text{ lb/day})}{(24 \text{ hr/day})(27.986 \text{ lb/lb mole})} = 4019.867 \text{ lb moles/hr}$$

Product Rates

C_2H_4: $(4019.867)(0.99) = 3979.668$ lb moles/hr

C_2H_6: $(4019.867)(0.004) = 16.080$ lb moles/hr

CH_4: $(4019.867)(0.006) = 24.119$ lb moles/hr

The required feed rate depends on the efficiency of the recovery system and ultimately one would have to correlate the design program for the reactor with that for the recovery system. For our purposes we will assume 92% recovery of ethylene in the product from the reactor and a 92% recovery of ethane from this same stream that is recycled. The recycle which will be the bottoms from an ethylene–ethane fractionator is estimated to contain 1.5% ethylene and 0.1% C_3's.

Using a once-through conversion of 60% the following material balance applies with discharge from furnace designated as D, recycle from recovery section as R, and fresh feed as F. The required ethylene is 3979.668 lb moles/hr.

$$\text{Ethylene in } D = \frac{3979.668}{0.92} \qquad\qquad = 4325.726 \text{ lb moles/hr}$$

$$\text{Ethane in } F + R = \frac{(4325.726)}{(0.6)(0.87)} \qquad = 8286.831 \text{ (based on selectivity of 87\% from Fig. 10.13)}$$

$$\text{Ethane in } D = 8286.831 - 4325.726 \quad = 3961.105$$

$$\text{Ethane in } R = (3961.105)(0.92) \qquad = 3644.217$$

$$\text{Ethylene in } R = (3644.217)\left(\frac{0.015}{1 - 0.016}\right) = \quad 55.552$$

$$\text{C}_3\text{'s in } R = (3644.217)\left(\frac{0.001}{1 - 0.016}\right) = \quad 3.704$$

$$\text{Total recycle} \qquad\qquad \overline{3703.473} \text{ lb moles/hr}$$

Total Feed to Furnace, lb moles/hr

Comp.	Fresh Feed	Recycle	Total Charge	Mole Fraction)	MW	(MW)(Mole Fraction)
CH$_4$	148.169*		148.169	0.0171	16.04	0.2743
C$_2$H$_4$		55.552	55.552	0.0064	28.05	0.1795
C$_2$H$_6$	4642.614	3644.217	8286.831	0.9589	30.07	28.8341
C$_3$'s	148.169	3.704	151.873	0.0176	44.09	0.7760
Total	4938.952	3703.473	8642.425	1.0000	MW =	30.06

$$* \ \frac{(4642.614)(0.03)}{0.94}$$

Design Variables

As outlined beginning on p. 435[1], we will set the following conditions:

Outlet Pressure. Use minimum permissable for flow to compressor suction, 10–15 psig.

Inlet Pressure. Based on estimated ΔP, 20 psi in the convection section and 30 psi in the radiant section, the inlet pressures are 62 psig for the furnace inlet and 42 psig for the radiant section inlet.

Outlet Hydrocarbon Partial Pressure and Steam Rate (see Table 10.6 p. 450[1]). Base calculations on 24 psia corresponding to 0.25 lb steam/lb hydrocarbon or $(0.25)(30/18) = 0.417$ moles steam/mole hydrocarbon.

Inlet Temperatures, Inlet to Furnace. The high-pressure feed will be reduced to the desired inlet pressure by using it first as a refrigerant in the distillation section, thereby conserving the high-pressure energy. The final temperature of this feed after multiple low-temperature refrigerant and cooling duties will be 60°F.

Inlet to Radiant (Reactor) Section. We are concerned here primarily with the radiant section where reaction occurs. This will begin in the region of reasonable incipient rates at 1250°F, as suggested on p. 454[1].

Tube Diameter. Tubes in the range of 4–6 in. will be considered. Thickness for centrifugal cast pipe for all sizes set at $\frac{3}{8}$ in.

Tube Lengths. Will vary with availability and furnace type. The following set lengths were selected for illustrative purposes.

	Tube Size, ID		
	4 in.	5 in.	6 in.
L_n, straight pipe length	32 ft	32 ft	32 ft
L, straight pipe plus length of 180° return bend	33.3 ft	33.3 ft	33.6 ft
L_B, equivalent length of return bend for ΔP, $L_B/D = 50$	16.7 ft	20.8 ft	25 ft
$L_n + L_B$, total equivalent length of pipe plus return bend	48.7 ft	52.8 ft	57 ft
$\omega_e = \dfrac{L_n + L_B}{L}$	1.463	1.586	1.696

Ethane Conversion Target. 60–65% (see p. 449[1]).

Mass Velocity. 22–25 lb/(ft²)(sec) is a typical value which provides reasonable heat transfer rates without excessive ΔP. Other values can be explored in order to compare various cases at constant ΔP. Studies at constant G_s, X_A, and P are essentially at constant u_s.

End-of-Run Maxima. $\frac{1}{16}$ in. coke thickness and 1800°F maximum tube wall temperature.

Heat Flux. Various values and patterns will be used. A typical average value for ethane cracking at high severity is 20,000 BTU/(hr)(ft²).

It becomes clear from this list that the major variables for study are tube diameter, heat flux, and mass velocity. The effects on yield and equilibrium approach, tube wall temperature, coil length, and ΔP will be noted in relation to allowable maxima and cost.

Models

Case 102A

Difference equations were used with Eq. 1.50, p. 30[1], as the heat balance with $q_0 \pi D_0 \Delta Z$ substituted for $q_z \Delta Z$, where q_0 is heat flux based on outside surface. The difference equation for conversion is

$$\Delta X_A = \frac{k_p P(1 - X_A)}{1 + n_{I_a} + \delta_A X_A} \frac{\Delta Z M_F}{G_s y_{A_0}} \qquad \text{(CS-2.3)}$$

where δ_A for ethane pyrolysis is 0.92 from Table 10.7 and k_p is determined from Eq. CS-2.2, corrected to time in seconds. The algorithm for computer calculation involved increments of one tube and return bend. The ΔP was calculated from Eq. 10.4 and then ΔX_A from Eq. CS-2.3 at inlet conditions. Composition of the mixture leaving the increment was then determined from equations of Table CS-2.1. The heat balance was then used to calculate outlet temperature. A new value of ΔX_A based on average of T_0 and T_e was then determined. Heat capacities and heats of formation were obtained from the *API Technical Data Book—Petroleum Refining*. Average values of the constants in the heat capacity polynomials for pseudocomponents, C_3's, C_4's, and C_5's are given in Table CS-2.3. Values for pure components are

Table CS-2.3   Heat Capacities and Heats of Formation

$$c_p^\circ = A \frac{T}{100} + B \left(\frac{T}{100}\right)^2 + C \left(\frac{T}{100}\right)^3 + D\left(\frac{100}{T}\right)$$

where $T = {}^\circ R$
Pure components: Data from *Technical Data Book—Petroleum Refining*, American Petroleum, Institute, New York, 1966.
Mixed components: Average values from same source based on typical compositions.

	MW	$A \times 10^2$	$B \times 10^3$	$C \times 10^5$	$D \times 10^1$	H_f° @ $77^\circ F$ BTU/lb
C_3's	43.08	8.208	−2.272	2.426	0.1620	−804.44
C_4's	56.62	8.213	−2.408	2.750	0.277	−145.54
C_5's	71.14	8.454	−2.476	2.810	0.004	−107.79

Table CS-2.4   Physical Properties

Viscosity[a]

$$\mu = \mu_r \mu_{cr}$$
$$\mu_{cr} = 7.70 M^{\frac{1}{2}} P_{cr}^{\frac{2}{3}} T_{cr}^{-\frac{1}{6}}$$
$$\ln \mu_r^{0.2} = (-0.1208 + 0.1354 \ln T_r)$$

where μ is the viscosity, micropoise, M is the molecular weight, P_{cr} is the critical pressure, atm, T_{cr} is the critical temperature, °K, and $T_r = T/T_{cr}$.

Thermal Conductivity

$$\text{Gaseous}^b: \quad \lambda_f = \frac{\mu c_v}{M}\left(\frac{3.670}{c_v} + 1.272\right)$$

where μ is the viscosity, lb/hr ft, c_v is the ideal gas heat capacity at constant volume, BTU/(lb-mole)(F), $(c_p - 1.99)$, and $\lambda_f = \text{BTU}/(\text{hr})(\text{ft})^2(°\text{F}/\text{ft})$

$$\text{Mean value}^c: \quad \bar{\lambda}_f = \frac{\sum y_j \lambda_j (M_j)^{\frac{1}{3}}}{\sum y_j M_j^{\frac{1}{3}}}$$

Metal: $\lambda_w = 14.1 + 0.00433\,(T - 1300)$, BTU/(hr)(ft²)(°F/ft) where: T is the temperature, °F

Coke: $\lambda_{ck} = 3.2$ BTU/(hr)(ft)(°F/ft), thickness $= \frac{1}{16}$ in.

[a] Ref. 8
[b] *Perry's Handbook.*
[c] *API Data Book.*

exactly as those appearing in the data book. The equilibrium approach was then calculated $(n_{C_2H_4} n_{H_2} P/n_{C_2H_6} K_{p_1})$, and the wall temperature using Eqs. 10.7 and 10.9, p. 437[1]. If in excess of 65% and 1800°F, respectively, the program was stopped. If not, the next increment was calculated.

Equations for physical properties needed in the calculations are given in Table CS-2.4.

Case 102B

Material balance equations in differential form for this model are summarized in Table CS-2.5. The differential heat balance Eq. 1.51, p. 30[1], was used with $q_0 \pi D_0$ substituted for q_z. Tube wall temperatures and equilibrium approach were calculated as in Case 102A at the end of each tube. Values of constants for the heat-capacity polynomials were obtained from pure component data presented in the *API Data Book*. This model was solved using a fourth-order Runge–Kutta method (11).

Table CS-2.5 Materials Balance Equations Case 102B

$$d\mathscr{F}_j = r_j \frac{\pi D^2}{4} dZ, \text{ where } \mathscr{F}_j = \text{moles of j/hr}$$

$$\frac{d\mathscr{F}_{C_2H_6}}{dZ} = (-r_1 - r_7)\frac{\pi D^2}{4}$$

$$\frac{d\mathscr{F}_{CH_4}}{dZ} = 2r_2 \frac{\pi D^2}{4}$$

$$\frac{d\mathscr{F}_{C_2H_4}}{dZ} = (r_1 - r_2 - r_3 - r_4 - r_5 - r_6 - r_7)\frac{\pi D^2}{4}$$

$$\frac{d\mathscr{F}_{C_3H_8}}{dZ} = (0.381r_7)\frac{\pi D^2}{4}$$

$$\frac{d\mathscr{F}_{C_3H_6}}{dZ} = 0.952r_7 \frac{\pi D^2}{4}$$

$$\frac{d\mathscr{F}_{C_2H_2}}{dZ} = r_5 \frac{\pi D^2}{4}$$

$$\frac{d\mathscr{F}_{H_2}}{dZ} = (r_1 - 2r_2 + 0.125r_3 + r_4 + r_5 + 2r_6 + 0.62r_7)\frac{\pi D^2}{4}$$

$$\frac{d\mathscr{F}_{C_4H_{10}}}{dZ} = 0.125r_3 \frac{\pi D^2}{4}$$

$$\frac{d\mathscr{F}_{C_4H_8}}{dZ} = 0.125r_3 \frac{\pi D^2}{4}$$

$$\frac{d\mathscr{F}_{C_4H_6}}{dZ} = 0.25r_3 \frac{\pi D^2}{4}$$

$$\frac{d\mathscr{F}_{C_6H_6}}{dZ} = 0.333r_4 \frac{\pi D^2}{4}$$

$$\frac{d\mathscr{F}_C}{dZ} = 2r_6 \frac{\pi D^2}{4}$$

Note. Rate expressions are given in Table CS-2.2.

Table CS-2.6   Material Balance Equations
Case 102C

$$\frac{d\mathcal{F}_{H_2}}{dZ} = r_4 \frac{\pi D^2}{4}$$

$$\frac{d\mathcal{F}_{CH_4}}{dZ} = r_2 \frac{\pi D^2}{4}$$

$$\frac{d\mathcal{F}_{C_2H_4}}{dZ} = (r_3 - r_6 - r_7) \frac{\pi D^2}{4}$$

$$\frac{d\mathcal{F}_{C_2H_6}}{dZ} = (-r_1 - r_2 - r_4) \frac{\pi D^2}{4}$$

$$\frac{d\mathcal{F}_{C_3H_6}}{dZ} = r_6 \frac{\pi D^2}{4}$$

$$\frac{d\mathcal{F}_{C_4H_{10}}}{dZ} = r_5 \frac{\pi D^2}{4}$$

Molar Flow Rate of Free Radicals[a]

$$\mathcal{F}_{CH_3\cdot} = \frac{2k_{c_1}\mathcal{F}_T RT}{k_{c_2} P} + \frac{k_{c_6}\mathcal{F}_{C_2H_4}}{k_{c_2}\mathcal{F}_{C_2H_6}}(\mathcal{F}_{C_2H_5\cdot})$$

$$\mathcal{F}_{H\cdot} = \frac{(\mathcal{F}_T RT/P)k_{c_3}}{k_{c_4}\mathcal{F}_{C_2H_6} + k_{c_7}\mathcal{F}_{C_2H_4}}(\mathcal{F}_{C_2H_5\cdot})$$

$$\mathcal{F}_{C_2H_5\cdot} = \sqrt{\frac{k_{c_1}RT\mathcal{F}_{C_2H_6}\mathcal{F}_T}{k_{c_5}P}}$$

[a] Obtained by solving the three simultaneous equations for constancy of free-
radical concentrations at the steady state.

Case 102C

This model was similar to Case 102B as was its solution. Material balances
are summarized in Table CS-2.6, and elementary steps with corresponding
rate constants in Fig. CS-2.1.

Design Procedure

The design procedure will be discussed using Case 102B. Comparison of the
final design selected using the models for all three cases will then be made.

Preliminary Calculations

Based on Eqs. 10.4, 10.7, 10.12, and 10.19 of Chapter 10, it is possible to converge rapidly upon the operating conditions required to maximize production of a given coil at the desired conversion while consuming the allowed ΔP and reaching the design tube wall temperature. The following relationships apply at constant values of G_s, X_A, and pressure

$$L\omega_e \propto D^{1.2} \tag{CS-2.5}$$

where $\omega_e = (L_n + L_B)/L$

$$q_0 \propto \frac{D^2}{LD_0} \tag{CS-2.6}*$$

$$L \propto \frac{1}{k_p} \tag{CS-2.7}$$

$$(T_w - T) \propto q_0 \frac{D_0}{D} \left[\frac{1}{h_i} + \frac{b_{ck}}{\lambda_{ck}} + \frac{b_w}{\lambda_w} \right]$$

or since the bracketed term does not vary greatly at constant G_s,

$$(T_w - T) \propto q_0 \frac{D_0}{D} \propto \frac{D}{L} \propto \frac{\omega_e}{D^{0.2}} \tag{CS-2.8}$$

Equation CS-2.8 is an approximation which indicates that in proceeding from 4 to 6-in. tubes, $T_w - T$ will increase only 7% or approximately 10–15°F, while at the same time the exit temperature will change by about the same order of magnitude. Thus as a first guess, one can assume T_w will remain roughly the same when Eqs. CS-2.5–CS-2.7 are applied.

The relationship between coil length and k_p, and thus temperature, can be based on the outlet temperature as shown in Eq. 10.19. It is thus possible to determine the length, average heat flux and heat-flux pattern, and outlet temperature that will produce the design ΔP, conversion, and tube wall temperature. As shown on p. 458[1], only one particular outlet temperature applies for a given tube size and this combination of variables. There is no need, however, to find this temperature with an accuracy greater than the original data warrant. A value within $\pm 5°$ is more than adequate. One set of operating conditions is selected by solving the design equations for one

---

* In coked condition D represents operating inside diameter based on an assumed coke thickness.

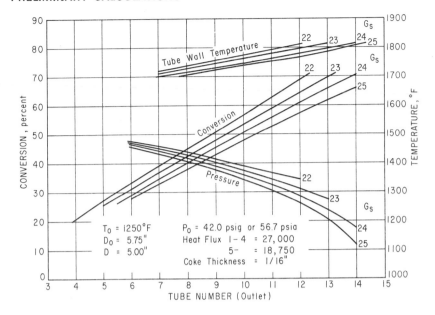

Fig. CS-2.2 Coil profiles for various mass velocities. (left y-axis also psia.)

tube size, as shown in Fig. CS-2.2. A mass flow rate of $G_s = 23$ is chosen as producing results closest to the design goals of 1800°F tube wall temperature, 60–65% conversion, and an outlet pressure of 30 psia. The corresponding values of variables for all other tube sizes can be rapidly determined by ratios based on Eqs. CS-2.5–CS-2.7, assuming a constant G_s and maintaining the same relative heat-flux pattern along the tube length. This pattern is selected so that approximately the initial one-third of the coil operates at 1.25 $\bar{q}_O$ and the remaining portion is 0.877 $\bar{q}_O$. A new L is obtained from Eq. CS-2.5 and then $\bar{q}_O$ from Eq. CS-2.6. The predicted outlet is then determined from Eq. CS-2.7.

These estimates can be checked by repeating the design calculations at the predicted values of q_O. Such calculated values are shown in Fig. CS-2.3. The solid lines represent the estimated values based on the 5-in. case. It is clear that the predicted values are within the accuracy of kinetic data. Actual calculated data for the three sizes are given in Table CS-2.7. The variations from design conditions are not significant since the nearest whole tube length must be used in any event. When more accurate data are available, the variation in selectivity should be included in the economic analysis. In the present case the accuracy of the data do not permit such analysis within the

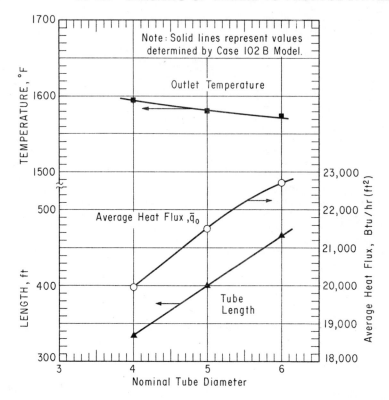

Fig. CS-2.3 Comparison of predicted and calculated values for 4-in. and 6-in. cases (based on 5-in. calculations).

apparent small variations in selectivity. Instead the design decision is made totally on the basis of tube costs and numbers of furnaces, assuming furnaces with four parallel coils. Based on a minimum of 7–12 furnaces (see p. 451[1]) and the apparent cost differences shown in Table CS-2.7, a 5-in. coil is selected with 8 furnaces. The capital costs shown for the radiant coils would exhibit even greater differences in the several furnace arrangements particularly when piping and maintenance are considered for the greater number of coils and furnaces required for the 4-in. case.

The design calculations and decisions thus far represent what might be termed the process design of the radiant coil. Design of the furnace necessary to provide the desired radiant heat flux together with the convection section for recovering heat from the high-temperature gases leaving the radiation section will not be considered (see p. 427[1]).

Table CS-2.7 Calculated Results Radiant Section

	Clean Tube ID, inches[a]		
	4	5	6
Conversion, %	64.66	64.66	65.09
Outlet temp., °F	1593	1581	1574
Tubewall temp., °F	1796	1795	1797
Outlet pressure, psig	16.0	16.5	16.6
Equilibrium approach, %	31.3	34.6	37.4
Avg heat flux, q_O = BTU/hr-ft^2	19,967	21,500	22,721
q_O for first portion of tubes shown in brackets	24,959[3]	27,000[4]	28,401[5]
q_O for remaining tubes as shown in brackets	17,828[7]	18,750[8]	19,565[9]
Ethane throughput/coil, lb/hr[b]	5426	8586	12,470
Total ethane feed, lb/hr required	←————————275,947————————→		
Coil length, ft	333.0	399.6	466.2
Tube weight/ethane feed throughput[c]	1.09	1.015	0.967
Total radiant coil cost, dollars, basis: $2.25/lb	676,760	630,194	600,391
Number of coils	51	32	22
Approximate number of furnaces	13	8	5–6

G_s = 23 lb/(ft)^2 sec.
Inlet pressure = 42 psig.
Inlet temperature: 1250°F.

[a] All tubes $\frac{3}{8}$ in. thick. Actual ID based on $\frac{1}{16}$ in. coke deposit.
[b] Basis: Ethane feed, as shown on p. 19 (96% ethane). Total flow of ethane feed plus steam is 1.25 times figure shown.
[c] Metal density = 496 lb/ft^3.

As described on p. 431[1], the convection section provides not only the source of preheat for the feed but also recovers excess energy that is used to preheat boiler feed water for the quench cooler and to superheat steam.

Design Comparisons for Three Alternates

Using the 5-in tube selected, design calculations based on the three models, Cases 102A, 102B, and 102C, are summarized in Fig. CS-2.4 and Table CS-2.8. The agreement for models A and B on conversions, temperatures

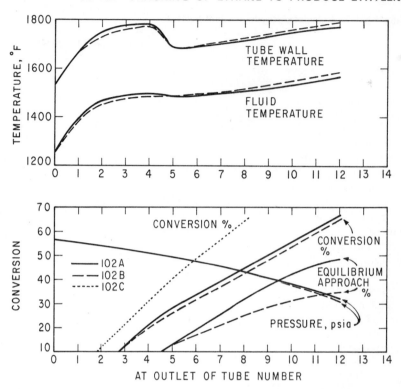

Fig. CS-2.4 Summary of profiles for 5-in. tube, cases A, B, and C (see Table CS-2.7 for conditions).

and ΔP is very good and that for compositions is good. Model C does not give good agreement. It reaches the desired outlet conversion at the outlet of the eighth tube rather than the twelfth. The data for this free-radical case are of excellent quality, but were based on studies not exceeding 20% conversion and in the absence of steam, with a quartz tube as a reactor. At higher conversions one would expect additional reactions involving ethylene disappearance that would have to be included in the model to make it agree with cases A and B. Cases A and B are based on both pilot and plant data and reproduce such data with good accuracy. Case C demonstrates the mode of handling free-radical mechanisms, but it also emphasizes, the hazards of using data not confirmed in pilot plants or operating plants. When data are obtained in the desired conversion range, the free-radical mechanism can be successfully applied, as emphasized on p. 457[1].

Table CS-2.8   Comparison of Calculated Product Distributions'

| Component | | Mole Fraction (Outlet, Tube No. 12 for A & B and Tube No. 8 for C) | | |
		Case A	Case B	Case C
Conversion, %		66.6	64.66	64.93
Hydrogen		0.3013	0.3004	0.2495
Methane		0.0329	0.0381	0.0667
Acetylene			0.0017	
Ethylene		0.2759	0.2399	0.2409
Ethane		0.1626	0.1702	0.1782
Propylene		0.0069	0.0129	0.0169
Propane			0.0141	0.0045
Butadiene			0.0028	
Butene		0.0046	0.0014	
Butane			0.0014	0.0227
Benzene	C_5's =	0.0042	0.0064	
Carbon			0.0013	
H_2O		0.2116	0.2094	0.2206
Total		1.0000	1.0000	1.0000

Basis: 5 in. coil; inlet conditions of radiant coil are 42 psig and 1250°F. $G_s = 23$ lb/ft^3 sec.

REFERENCES

1. P. D. Pacey and J. H. Purnell, *Ind. Eng. Chem. Fundam.*, **11**, 233 (1972).
2. D. A. Leathard and J. H. Purnell, *Ann. Rev. Phy. Chem.*, **21**, 197 (1970).
3. B. L. Crynes and L. F. Albright, *Ind. Eng. Chem. Process Des. & Develop.*, **8**, 25 (1969).
4. C. P. Quinn, *Trans. Faraday Soc.*, **59**, 2543 (1963).
5. H. R. Linden and R. L. Peck, *Ind. Eng. Chem.*, **47**, 2470 (1955).
6. H. C. Schutt, *Chem. Eng. Prog.*, **55** (1), 68 (1959).
7. P. S. Myers and K. M. Watson, *Nat. Petrol. News, Tech. Sec.*, **38**, R 388 (May 1, 1946) and R 439 (June 5, 1946).
8. R. H. Snow and H. C. Schutt, *Chem. Eng. Prog.*, **53**, 133M (1957).
9. M. J. Shah, *Ind. Engr. Chem.*, **59** (5), 70 (1967).
10. I. Lichtenstein, *Chem. Eng. Prog.*, **60** (12), 69 (1964).
11. G. E. R. Franks, *Modeling and Simulation in Chemical Engineering*, Wiley-Interscience, New York, 1972.
12. J. R. Fair and H. F. Rase, *Chem. Eng. Prog.*, **50**, 415 (1954).

CASE STUDY 103

Quench Cooler

THIS STUDY illustrates a quench design problem that is applicable to any very fast reaction in which reaction continues during the quenching process. In this particular case the reaction must be quenched as rapidly as possible, while at the same time minimizing pressure drop, since the ultimate destination of the effluent is a compressor suction.

Problem Statement

Design a quench cooler for quenching the effluent from two coils of the ethane cracking furnace of Case 102B.

Chemistry and Kinetics

The chemistry and kinetics are already documented in Case Study 102. For this quench cooler the homogeneous reaction continues to occur in the gases discharging from the furnace. If the model accurately described the coke-forming reaction both homogeneous and heterogeneous, it would be more valuable in the design of the quench cooler, for it would then provide a means for determining on stream time for the cooler by monitoring the changes in temperature profile and pressure as coke is deposited on the tube walls.

The model of Case 102B will be used. Although it will correctly predict the formation of coke at the inlet end of the exchanger for ethane cracking effluent, the quantitative data are not sufficiently accurate to use seriously in any operating time study. The heat effects, however, are sufficiently accurate for describing the declining reaction as required for determining heat transfer area.

Models for quench cooler fouling exist but details have not been presented (1).

Design Basis

For design purposes the exit conditions for Case 102B will be used with the total amount flowing equivalent to that from two 5-in coils. Outlet temperature should be safely above the dew point of the mixture which depends primarily upon the small amounts of heavy polymeric products not normally included in the reaction model. In ethane cracking the amount of $400°F$ plus material is so small that the dew point is very low, and the outlet temperature and thus the steam temperature is based solely on the steam economics in the particular plant (2). The decision would also be based in part on the steam turbines to be selected. Higher pressure steam produces better turbine efficiency (lower water rates). For illustrative purposes we select 1500 psia because of the number of large compressors in the operating area. If the main use for this system would be smaller turbines, 900 psia would be a better pressure level at this point in time since turbines at this level are cheaper and are produced by most manufacturers (5). The following design parameters will be used.

Inlet Conditions. $1581°F$ @ 31.2 psia.

Composition.   See Table CS-2.8 Case 102B, p. 31.

Steam Pressure.   1500 psia @ $596°F$ (prior to use, steam will be superheated.)

Double-Pipe Tube Properties.   Inside tube: 0.97 in. ID and 1.25 in. OD. Outside tube and materials of construction: to be specified by manufacturer.

Maximum Allowable ΔP. ≈ 2 psi.

Maximum Tube Length.   Based on given limitations of the correlation for the film coefficient a maximum L/D of approximately 240 will be set. This limits the tube length to approximately 20 ft.

Design Equations

The same heat and material balance and pressure-drop equations as those given for Case 102B are used. In Eq. 1.50 the heat-flux term becomes

$$q_O = U(T - T_j) \tag{CS-3.1}$$

where U is the overall film coefficient, T is the gas temperature, °F, and T_j is the boiling water temperature, °F. For the double-pipe design, shown in Fig. 10.10,

$$U = \frac{D_O}{Dh_i} + \frac{D_O b_w}{D_m \lambda_w} + \frac{1}{h_o} \tag{CS-3.2}$$

where h_i is the gas film coefficient and h_o is the boiling water coefficient taken as 2040 BTU/(hr)(ft²)(°F).

The high flux encountered in such exchangers creates steep temperature gradients and distorted velocity profiles that demand special correlations (3). The following equation for h_i has been recommended based on a thorough review of the literature (4).

$$\left(\frac{h_i D}{\lambda_f}\right)_Z = \frac{0.021(N_{Re})_Z^{0.8}(N_{Pr})_Z^{0.4}}{(T_w/T)^{0.29+0.0019Z/D}} \tag{CS-3.3}$$

where T and T_w are the bulk fluid and the wall temperatures, respectively, and sub Z refers to the value of the parameter at Z distance from entrance.

This correlation gives results within 10% for most high temperature data (3, 4) in the following ranges:

$$10 < \frac{L}{D} < 240$$

$$200 < T < 2800°R$$

$$1.1 < \frac{T_w}{T} < 8.0$$

Design Procedure

Various values of mass velocity will be tried and results will be compared on the basis of heat recovered and pressure drop. Calculations are made for one tube, and then the exchanger is designed so that the total flow will be distributed equally among all tubes.

Results

In Table CS-3.1 the results of several calculations are given. It is clear that the case for $G_s = 12$ meets the design requirements. Although the additional conversion in the exchanger tube is small, it does increase with lower mass velocities. Conversion at the low average temperature of the exchanger does produce undesired side reaction products including some which are not very precisely documented by this model. It would seem wise, therefore, to use the maximum allowable ΔP and minimize thereby the extent of conversion.

Table CS-3.1 Comparison of Design Cases Quench Cooler

Mass Velocity lb/ft² sec	ΔP, psi	Outlet Temp. °F	Conversion
10	1.48	670	2.59×10^{-3}
12	2.1	679	2.59×10^{-3}
13	2.44	683	2.17×10^{-3}

From Table CS-2.7, p. 29, the total output for two coils is

$$(8586)(2)(1.25) = 21465 \text{ lb/hr}$$

Number of exchanger tubes $= 21465/[(3600)(12)(\pi/4)(0.97/12)^2] = 96.8$ or 97 tubes.

REFERENCES

1. J. Chen and M. J. Maddock, *Hydrocarbon Process.*, **52** (5), 147 (1971).
2. J. P. Fanaritis and H. J. Streich, *Chem. Eng. (N.Y.)*, p. 80 (May 28, 1973).
3. D. W. Sundstrom and R. L. DeMichell, *Ind. Eng. Chem. Process Des. & Develop.*, **10**, 114 (1971).
4. W. R. Gambil, *Chem. Eng. (N.Y.)*, **74** (18), 147 (1967).
5. J. F. Farrow, *Hydrocarbon Process.*, **50** (3), 71 (1971).

CASE STUDY 104

Toluene Dealkylation

THIS STUDY illustrates a case where pilot-plant data already indicate the temperature range for practical rates. Design calculations are made in this range in order to select a design basis.

Problem Statement

Design a noncatalytic reactor for producing 82,000 gal/day (25,000 lb/hr) of benzene by thermal hydrodealkylation of toluene. Hydrogen is available from another unit at 600 psig and 80°F. Toluene feed will normally be essentially pure but can be contaminated with paraffinic hydrocarbons at times.

Chemistry

A number of possible reactions are reported in addition to the primary reaction (1).

$$C_6H_5CH_3 + H_2 \rightleftharpoons C_6H_6 + CH_4$$

These include hydrogenation of benzene and toluene to cycloparaffins, hydrocracking of paraffins, and reaction of benzene and toluene to various diphenyls.

A free-radical mechanism has been suggested (2).

$$H_2 \rightleftharpoons 2H\cdot$$
$$H\cdot + C_6H_5CH_3 \longrightarrow CH_4 + C_6H_5\cdot$$
$$H\cdot + C_6H_5CH_3 \longrightarrow C_6H_6 + CH_3\cdot$$
$$CH_3\cdot + H_2 \longrightarrow CH_4 + H\cdot$$

$$C_6H_5 \cdot + H_2 \longrightarrow C_6H_6 + H \cdot$$
$$C_6H_5 \cdot + C_6H_6 \longrightarrow C_{12}H_{10} + H \cdot$$
$$C_6H_5 \cdot + C_6H_5CH_3 \longrightarrow C_{12}H_{10} + CH_3 \cdot$$
$$C_6H_5 \cdot + CH_4 \longrightarrow C_6H_5CH_3 + H \cdot$$

Saturated hydrocarbons if present are thermodynamically favored to hydrocrack to CH_4 and those of C_6 and above do so very rapidly with high heat effects. Success of the process depends on preventing hydrogenation of benzene to cyclohexane and high heat effects of hydrocracking which could occur with saturated impurities or cyclohexane produced in the reactor. Cyclohexane if not hydrocracked will contaminate the benzene product because of its close boiling point to benzene (1).

Thermodynamics

Heats of reaction for the main reaction and heat capacity data are plotted in Fig. CS-4.1 and CS-4.2 and heats of reaction at 800°K for several overall reactions are tabulated in Table CS-4.1. Equilibrium constraints for major overall reactions are plotted in Fig. CS-4.3. Clearly, temperatures above 800°F are required to prevent hydrogenation of toluene and benzene. Although the heat of reaction of toluene dealkylation is modest, much larger heat effects accompany hydrocracking of cyclohexane or saturated hydrocarbons. It is known that these hydrocracking reactions are very fast.

Because of the low value of the equilibrium constants for the side reactions, high temperatures should not encourage forming excessive amounts of diphenyl, methyl diphenyl, and dimethyl diphenyl, and the equilibrium amounts can be reduced by increasing the hydrogen partial pressure (3,4).

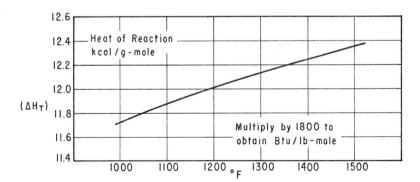

Fig. CS-4.1   Heat of reaction for toluene dealkylation.

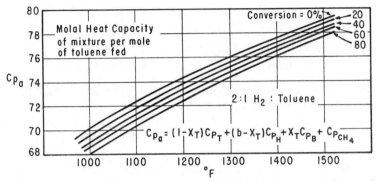

For other H_2 : toluene ratios (b) add $(7.2)(b-2)$ to values of C_{P_a} read.

Fig. CS-4.2 Heat capacity for toluene dealkylation.

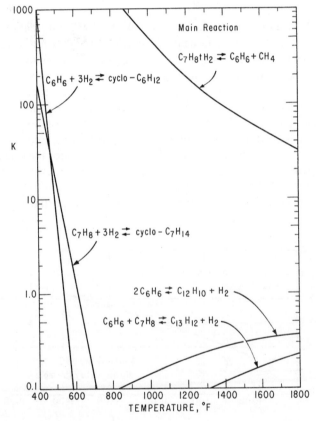

Fig. CS-4.3 Equilibrium constants for various reactions in toluene dealkylation.

Table CS-4.1   Heats of Reaction of Typical Overall Reactions

	ΔH 800°K kcal/g mole
1. $C_6H_5CH_3 + H_2 \rightarrow C_6H_6 + CH_4$	−11.71
2. $C_6H_6 + 3H_2 \rightleftarrows$ cyclo C_6H_{12}	−52.70
3. Cyclo $C_6H_{12} + 6H_2 \rightarrow 6CH_4$	−87.75
4. $nC_6H_{14} + 5H_2 \rightarrow 6CH_4$	−76.66
5. $CH_4 \rightarrow C + 2H_2$	+22.82

Although the equilibrium constant for the main reaction declines with increasing temperature, its value is such that benzene is still strongly favored at the highest practical temperatures.

The gases behave ideally as determined by pseudocritical calculations. Both fugacity coefficients and compressibilities have values of unity.

Kinetics

The first several steps of the free-radical mechanism can be rationalized in an empirical form for the disappearance of toluene (1,5).

$$(-r_{To}) = (k_{c_1} + k_{c_2})[H\cdot][C_7H_8]$$

$$K_1 = \frac{[H\cdot]^2}{[H_2]} \quad \text{or} \quad [H\cdot] = [H_2]^{0.5}[K_1]^{0.5}$$

Thus

$$(-r_{To}) = k_c C_{To} C_{H_2}^{0.5}, \frac{\text{(lb moles toluene)}}{(\text{ft}^3)(\text{sec})}$$

where C_{To} is the concentration of toluene and C_{H_2} is the concentration of H_2.

Rate data in the range of commercial interest yield the following (6):

$$k_c = 7.18 \times 10^{11} e^{-98.100/R'T}, (\text{ft}^3)^{0.5}/(\text{lb mole})^{0.5}(\text{sec})$$

or

$$k_p = \frac{k_c}{(RT)^{1.5}}, (\text{lb mole})/(\text{atm})^{1.5}(\text{ft}^3)(\text{sec}) \qquad \text{(CS-4.1)}$$

$$(-r_{To}) = k_p P_{To} P_{H_2}^{0.5}$$

$$R' = 1.987$$

$$R = 0.73, (\text{ft}^3)(\text{atm})/(\text{lb mole})(°R)$$

$$T = °R$$

Rate equations for the side reactions have not been reported, but pilot-plant results at 1300°F and 600 psig indicate that equilibrium is attained for the various diphenyls. The total amount of by-products formed is inversely proportional to the H_2-to-toluene ratios (7). Selectivity to benzene is 98% at conversions up to 75% and H_2-to-toluene ratios of at least 2 : 1 (3,4).

Because of the apparent equilibrium for diphenyls, it will be advantageous to recycle diphenyls to reduce the net consumption of toluene and benzene to these side-reaction products.

Reactor Type

Since the adiabatic factor and heat generation potential are modest (see Table 6.4), an adiabatic reactor will be selected. Because of the possibility of paraffinic contamination, however, inlets should be installed along the reactor for quenching temperature excursions caused by hydrocracking of paraffins. A plug-flow reactor will require the smallest volume.

Operating Conditions

Pressure, temperature, and conversions are selected on the basis of thermodynamic and kinetic data together with pilot-plant data.

Pressure

Equilibrium conversions of the major reactions are unaffected by pressure, but pressure will increase the rate and reduce equipment size. Since the hydrogen is available at 600 psig, a pressure of 50 psi less, for furnace and other AP ahead of the reactor, will be selected (550 psig).

Conversion and H_2-to-Toluene

A conversion of 75%, 98% selectivity, and a H_2-to-toluene ratio (b) of 2 are selected as consistent with pilot-plant data.

Temperature

Pilot-plant data (3,4) suggest operating temperatures in the range of 1150–1350°F. This range can be investigated. In this range with hydrogen present carbon steel and most low and intermediate alloy steels cannot be used (See Fig. B.1, Appendix B). A refractory-lined reactor should be specified with an inner protective liner of stainless steel. A maximum design temperature of 1500°F will thus be set with something in the neighborhood of 1400°F being a normal operating limit.

Reactor Design

In the temperature range of interest the energy balance based on Figs. CS-4.2 and CS-4.3 is well represented as follows:

$$\Delta T = \frac{(-\Delta H_{To})(\Delta X_{To})}{c_{p_a}} \approx \frac{(12.08)(1800)\Delta X_{To}}{74.5}$$

$$\Delta T \approx 292 X_{To} \qquad\qquad (\text{CS-4.2})$$

Toluene Balance

$$\frac{dV}{F_{To}} = \frac{dX_{To}}{k_p P^{1.5}\left(\dfrac{1 - X_{To}}{1 + b}\right)\left(\dfrac{b - X_{To}}{1 + b}\right)^{0.5}} \qquad (\text{CS-4.3})$$

where b is H_2-to-toluene ratio, X_{To} is the conversion of toluene, F_{To} is the feed rate of toluene, lb moles/sec.

$$F_{To} = \frac{25{,}000 \text{ lb/hr}}{(3600)(78)(0.98)(0.75)} = 0.121 \text{ lb moles/sec @ } 75\% \text{ conversion}$$

For rapid estimates on this system analytically integrate Eq. CS-4.3 using k_p evaluated at $(1/T) = \frac{1}{2}[(1/T_0) + (1/T_e)]$.

Since for 75% conversion $\Delta T = (292)(0.75) = 219$, the maximum inlet design temperature should be $1400 - 219 = 1181$ or approximately $1200°F$. Cases at 1150 and 1200°F were calculated and are summarized in Table CS-4.2. With reference to this table it is seen that pressure fluctuations do not have much effect. It is not practical to use L/D values of 100, for the length would be too great. Since the 3 ft–0 in. ID size allows more room for installing and repairing refractory lining, it will be selected.

Table CS-4.2 Summary Dealkylation of Toluene

Inlet Conditions		Outlet Temp. °F	Volume cu ft	Total Length	
Temp. °F	Press. psig			2 ft diam.	3 ft diam.
1. 1150	550	1369	608	195	86
2. 1200	550	1418	275	88	39
3. 1150	575	1369	569	181	80

H_2-to-Toluene = 2. Conversion = 75%. Based on numerical solution of the simultaneous balances for mass and heat with the latter in terms of $c_p = f(T, X_{To})$.

Because of low values of L/D some longitudinal dispersion will occur. Hence select Case 1 at 1150°F inlet. This will allow adequate flexibility for raising temperature, if necessary, to meet production requirements.

Final Design

Two vertical reactors, 3 ft–0 in. ID × 43 ft connected in series and lined with refractory with protective type 347 stainless-steel inner liner. Outside shell shall be a low alloy steel for protection against leaks in refractory. Paint with heat sensitive paint, and insulate outside portion only where personnel protection requires. Install quench inlets along length at 5 ft, 10 ft, and 30 ft, respectively of each.

Pressure Drop

The pressure drop in the reactor sections will be negligible, based on Fig. 10.3 on p. 424[1], and a Reynolds number in the turbulent region of

$$N_{Re} = \frac{(6.3)(41{,}818)}{(36)(0.03)} = 243{,}938$$

Hydrogen recycle compressor operating costs will thus be minimized.

Estimate of Effect of Dispersion on Conversion

N_{Re} is in turbulent region. Thus $(N_{Pe})_a = 4$. As an approximation use effective $L/D = 43/3 \approx 14$.

From Eq. 10.35, p. 486[1],

$$(X_A)_{plug} - (X_A)_{actual} = \frac{1}{(N_{Pe})_a} \frac{D}{L} \frac{V}{F_A} (-r_A)_{plug} \ln \frac{(-r_A)_{plug}}{(-r_A)_{0plug}}$$

$$\Delta X_A = \left(\frac{1}{4}\right)\left(\frac{1}{14}\right)\left(\frac{608}{0.121}\right)(3.59 \times 10^{-4}) \ln \frac{3.59 \times 10^{-4}}{5.527 \times 10^{-5}}$$

$$\Delta X_A = 0.06, \quad \text{negligible.}$$

REFERENCES

1. G. F. Asselin, in *Advances Petroleum Chemistry and Refining*, Vol. 9, Interscience, New York, 1964, pp. 47–97.
2. G. J. F. Stigntjes, H. Voetter, E. F. Roelofsen, and J. J. Verstappen, *Erdol Kohle*, **14**, 1011 (1961).

3. S. Feigelman, L. M. Lehman, F. Aristoff, and P. M. Pitts, *Hydrocarbon Process*, **44** (12), 147 (1965).
4. M. J. Fowle and P. M. Pitts, *Chem. Eng. Prog.*, **58** (4), 37 (1962).
5. R. I. Silsby and E. W. Sawyer, *J. Appl. Chem.*, **6**, 347 (1956).
6. S. E. Shull and A. N. Hixon, *Ind. Eng. Chem. Process Des. Develop.*, **5**, 147 (1966).
7. C. C. Zimmerman and R. York, *Ind. Eng. Chem. Process Des. Develop.*, **3**, 255 (1964).

CASE STUDY 105

Shift Conversion

THIS STUDY illustrates the design of massive adiabatic, catalytic reactors and the use of thermodynamics to set operating conditions. Development of necessary details for a process vessel sketch preliminary to detailed mechanical design is demonstrated.

Problem Statement

Design a shift conversion system for converting the carbon monoxide present in the gas from a reformer of an ammonia plant to carbon dioxide. The following specifications apply. Figure CS-5.1 is a flow diagram for an ammonia plant showing the three major reactors—reformer, shift converter, and ammonia converter.

Feed and Product Specifications

Synthesis gas from the reforming section of an ammonia plant is available at the following conditions:

Feed (Dry Gas Basis): 12,400 lb moles/hr

	Mole %	MW	(%)(MW)
CO	13.0	28	364.0
CO_2	7.9	44	347.6
H_2	56.8	2	113.6
N_2	21.8	28	610.4
CH_4	0.5*	16	8.0
	100.0		1443.6 $M_m = 14.44$

* Argon included in methane.

44

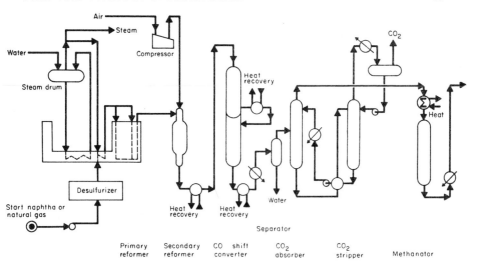

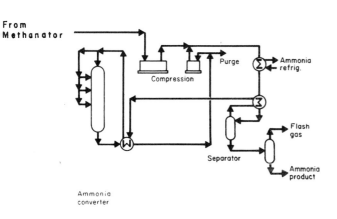

Fig. CS-5.1· Schematic flow diagram for an ammonia complex. Courtesy: The Pullman Kellogg, Division of Pullman, Inc., Houston, Tex.

Total wet feed from reformer: 20,460 lb moles/hr, $M_m = 15.84$.

Reformer discharge conditions: 400 psia, 1750°F.

Impurity specifications: CO—0.2% to 0.5% after shift conversion (dry gas basis) is a typical range. Value selected depends on steam-to-CO ratio and other factors.

Fuel cost basis: \$0.50/MM BTU.

Catalysts

Two catalysts are available for the shift conversion, chromia-promoted iron catalyst, which has been used many years for the shift reaction, and a copper-zinc catalyst that offers the thermodynamic advantage of lower operating temperatures for this exothermic reaction. Additional catalyst data are given in the following:

Type	Chromia-promoted iron oxide	copper-zinc oxide
Maximum operating temp., °F	890	500–550
Tablet size (in.)	$\frac{1}{4} \times \frac{1}{4}$	$\frac{1}{4} \times \frac{1}{8}$
Bulk density (lb/cu ft)	70	90
Particle density (lb/cu ft)	126	155
Cost ($/cu ft)	20.00	75.00
Catalyst poisons	Inorganic salts, boron, oils, or phosphorous compounds, liquid H_2O is a temporary poison. Sulfur compounds in an amount greater than 50 ppm	Sulfur and halogen compounds, and un-saturated hydro-carbons
Catalyst life	3 yr and over depending on care in startup and opera-tion (Use times up to 15 yr have been reported)	2–3 yr

Chemistry and Kinetics

Overall Reaction: $CO + H_2O \rightleftarrows CO_2 + H_2$

The iron-oxide catalyst has been studied in greater detail [1]. CO is apparently chemisorbed on the iron oxide and both water and CO_2 are strongly adsorbed. CO_2 has an inhibiting effect on the forward rate, and hydrogen appears not to be adsorbed. There are no significant side reactions.

Various manufacturers suggest rate equations for their catalysts. One such manufacturer recommends the following which will be assumed to represent midlife activity (2):

$$(-r_{CO}) = \psi k(y_{CO}y_{H_2O} - y_{CO_2}y_{H_2}/K)/(379\rho_b) \qquad \text{(CS-5.1)}$$

where k = rate constant
$\quad$ = $\exp(15.95 - 8820/T)$ for iron catalyst
$\quad$ = $\exp(12.88 - 3340/T)$ for copper-zinc catalyst
$\quad K$ = equilibrium constant
$\quad$ = $\exp(-4.72 + 8640/T)$ for $760 \le T \le 1060$
$\quad$ = $\exp(-4.33 + 8240/T)$ for $1060 \le T \le 1360$
$\quad P$ = pressure, atm
$(-r_{CO})$ = rate, lb moles CO converted/(lb catalyst)(hr)
$\quad T$ = temperature, °R
$\quad y_j$ = mole fraction of component indicated
$\quad \rho_b$ = catalyst bulk density, lb/cu ft
$\quad \psi$ = activity factor

Iron catalyst ψ = $0.816 + 0.184P$ for $P \le 11.8$
$\qquad\qquad\quad$ = $1.53 + 0.123P$ for $11.8 < P \le 20.0$
$\qquad\qquad\quad$ = 4.0 for $P > 20.0$

Copper-zinc catalyst ψ = $0.86 + 0.14P$ for $P \le 24.8$
$\qquad\qquad\qquad\quad$ = 4.33 for $P > 24.8$

Anyone seeking confirming evidence of the efficacy of rate equations in predicting plant performance will certainly find that equations recommended by catalyst manufacturers have been subjected to numerous tests, including many observations on full-scale plants. The rate constants are expressed on the basis of a reasonable "lined-out" activity that the catalyst would maintain for a considerable time provided operating errors which cause deactivation do not occur. We will assume that this equation represents activity levels characteristic of mid-life of the catalyst.

Extensive investigations on the iron catalyst, as discussed on p. 150[1], confirm the form of Eq. CS-5.1 and the ψ term is shown to be the product of the total pressure in atmospheres and ratio of the first-order constant at pressure P to that at atmospheric pressure and is a function of pressure and the Thiele modulus, as shown in Eq. 3.47, p. 153[1]. The manufacturer's recommended equation for ψ closely approximates values obtained from Eq. 3.47.

Thermodynamics

Equilibrium CO mole fraction and adiabatic reaction temperature plots are presented in Fig. CS-5.2 based on the following equations.

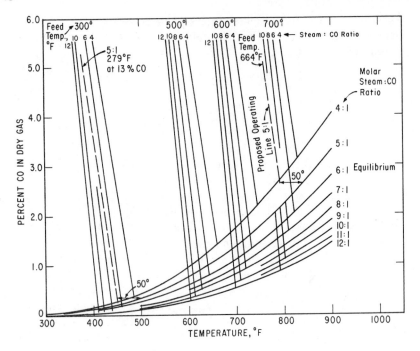

Fig. CS-5.2   Adiabatic plot for shift conversion at various steam-to-CO ratios.

Basis: 100 moles dry gas

$$K = \frac{n_{CO_2}n_{H_2}}{n_{CO}n_{H_2O}} \quad K_v = \frac{(7.9 + \Delta n_{CO})(56.8 + \Delta n_{CO})}{(13.0 - \Delta n_{CO})(13.0b - \Delta n_{CO})} \quad (\text{CS-5.2})$$

where b is the moles of steam per mole CO and n is the indicated moles per 100 moles of dry gas.

The fugacity correction term $K_v \approx 1$ for this system.

$$\text{Percentage of CO in dry gas} = \frac{13.0 - \Delta n_{CO}}{100 + \Delta n_{CO}} \times 100$$

Energy Balance

Basis: 25°C and the elements; 100 moles of feed

$$\sum (n_j H_{f_j}°)_F + \sum \left(n_j \int_{298}^{T_F} c_{p_j} dT \right)_F = \sum (n_j H_{f_j}°)_P + \sum \left(n_j \int_{298}^{T} c_{p_j} dT \right)_P$$

where F refers to feed and P to product.

Ideal molal gas heat capacity equations are adequate since for this system $c_p - c_p^*$ is negligible. The following equation and constants were used (4).

$$c_{p_j} = a + bT + cT^2 + dT^3, \quad T = {}^\circ K \qquad \text{(CS-5.3)}$$

	a	$b \times 10^2$	$c \times 10^5$	$d \times 10^9$
CO	6.726	+0.04001	+0.1283	−0.5307
H_2O	7.7	+0.04594	+0.2521	−0.8587
CO_2	5.316	+1.4285	−0.8362	+1.784
H_2	6.952	−0.04567	+0.09563	—0.2079
N_2	6.903	−0.03753	+0.1930	−0.6861
CH_4	4.75	+1.2	+0.303	—2.63

Quite obviously an adiabatic reactor is possible, for the changes in temperature are modest particularly at high steam ratios which are needed to improve the equilibrium because of the large amount of hydrogen in the feed. The heat generation potential is also quite low (1.12, see Table 6.4). The steam minimizes temperature change by increasing the heat capacity of the system.

It is clear from Fig. CS-5.2 that when assuming an optimum equilibrium approach temperature of 50°, as recommended for this system, the desired reduction in CO cannot be attained with the iron catalyst when operating in the range of 750°F (100° less than maximum allowable). By operating at 400°F with the copper-zinc catalyst (100° less than its maximum allowable) it is possible to attain the desired reduction in CO, but this catalyst is more expensive. Three alternates are possible.

1. Remove part of CO with iron catalyst in one bed. Then absorb CO_2 and go to a second bed of the same catalyst with a more favorable equilibrium since the product CO_2 is absent.

2. Conduct the entire reaction in a single bed on zinc-copper catalyst.

3. Remove part of the CO in a bed with iron catalyst and complete the removal in a second bed of the more expensive copper-zinc catalyst.

The second and third alternates will be considered as the most attractive if for no other reason than additional absorption equipment creates added maintenance problems particularly because of the corrosive character of monoethanolamine, the usual absorbent.

Design Conditions

General decisions on operating conditions are now possible.

Pressure

Since pressure increases the reaction rate, the shift convertor should operate at the reformer pressure less the drop through the waste-heat boiler after the secondary reformer and associated piping (~ 10 psi). Pressures of 390 psia for first reactor and 380 psia for second will be used.

Steam-to-CO Ratio

Steam-to-CO ratios must surely be in the range of 4 : 1 or higher as indicated by Fig. CS-5.2. The optimum value can only be determined by economic analysis based on design studies. On the negative side higher steam rates cause greater flows and larger diameter equipment. Since steam is also required for the secondary reformer (Fig. CS-5.1), there is substantial logic in adding all needed steam at that point. The recoverable heat is in a more valuable form, and the reformer equilibria are also favorably influenced. Alternatively, additional steam might be injected as quench for the secondary reformer in lieu of a waste heat boiler. This particular case study would normally be conducted as part of a general design study for the entire ammonia plant, and the effects of changes on the total economics assessed. The feed stream given has a ratio of 5:1 steam-to-CO, which is quite adequate for the shift reaction. Referring to Fig. CS-5.2 even a 12:1 ratio at 500°F feed temperature will not accomplish the desired removal in one stage with the high-temperature catalyst.

If an upper limit of 450°F is to be maintained for the low temperature catalyst, it is clear from Fig. CS-5.2 that the second alternate, a system with all low-temperature catalyst is not feasible. The required inlet temperature of 279°F would be below the dew point of the feed (see the following). Thus only alternate 3 remains to be considered.

Temperature and Outlet CO Concentration

A minimum inlet temperature must be established relative to the dew point. Contact of either catalyst with liquid water at operating temperatures will cause thermal shock and disintegrate the catalyst. An approximate dew point may be estimated as follows: First converter:

$$\text{Partial pressure of steam} = \frac{20{,}460 - 12{,}400}{20{,}460}(390)$$

$$= 154 \text{ psi}$$

Corresponding temperature of saturated steam is 361°F. First converter inlet will be substantially above this value. Assume for purposes of this estimate that 85 % of CO will be removed in first reactor. Hence

$$\text{Inlet second converter} = \frac{20,460 - (12,400)[1 + (0.13)(0.85)]}{20,460}(380) \quad (380)$$

$$= 127 \text{ psi}$$

Corresponding saturated steam temperature is 345°F. Therefore, design for minimum inlet temperature of around 25° above 345°F, approximately 370°F at inlet of secondary reactor.

The maximum temperatures of 890 and 550°F, respectively, for the two catalysts should be used for vessel mechanical design, but process design should be based on a lower temperature. Thus the outlet temperatures must be less than these values and such that the desired residual CO can be reached.

The manufacturer recommends a 50° approach to equilibrium. Based on Eq. 11.37, p. 528[1], the value is 100°F, but the manufacturer's suggestion will be used, and cases will be selected on both sides of this value to minimize the catalyst requirements.

Referring to Fig. CS-5.2 in the low temperature region, it is clear that based on a 50° approach and 5 : 1 steam-to-CO ratio that catalyst performance at mid-life of 0.35 % CO in the outlet would require a 450°F outlet. This seems a reasonable temperature around which to structure cases since it is sufficiently removed from the maximum allowable of 500–550°F. By following an adiabatic line back to 400°F we see that a reasonable inlet range (outlet from first reactor) is 3.5 % or less. Using a value of 3.0 % with a 50° approach brings us to an 820°F outlet for the first reactor. It seems reasonable to study values at 3.0 % and below, since higher values will leave little safety factor from the 890°F and place the inlet to reactor no.2 closer to the dew point region.

Referring to the equations for the rate constants for both catalysts, it can be shown that the rate doubles for a rise of 100° for the iron catalyst and 200° for the copper-zinc catalyst. Thus using 790°F as an approximate maximum for design, one has the opportunity to raise the temperature to compensate for a 50 % loss in activity for the iron catalyst. In the case of the low-temperature catalyst it is not possible to set the outlet temperature much less than 450°F.

Since the rate data are based on activities at mid-life, the unit will perform better than design at the outset and can be kept at design outlet CO values by altering temperature strategy as the activity declines toward the last half of its life. Toward the end of this period, however, temperature increases may no longer be effective because the approach to equilibrium will become too close and higher outlet CO concentrations will have to be accepted.

Design Cases

Two-Stage System (5 : 1, steam: CO):

1. Calculate minimum catalyst requirements for the following cases which were selected in the range of the outlet CO previously specified.

CO Outlet Conc. %	Inlet Temp., °F
2.25	600, 610, 620, 630, 640, 650, 670
2.65	630, 640, 650, 664, 670, 680, 700
3.0	650, 660, 670, 680, 700, 710, 720

2. Calculate the second stage for convenience starting at 3.0% and ending at 0.2%, so that intermediate inlet and outlet values may be selected as desired.

Inlet temperatures: 370 and above.

Design Procedure

Component mole balance

$$\Delta W(-r_{CO}) = (-\Delta \mathscr{F}_{CO}) \qquad \text{(CS-5.4)}$$

$$\mathscr{F}_{CO_{i+1}} = \mathscr{F}_{CO_i} - (-\Delta \mathscr{F}_{CO}); \ \mathscr{F}_{CO_{2_{i+1}}} = \mathscr{F}_{CO_{2_i}} + \Delta \mathscr{F}_{CO_2}, \text{ etc.}$$

where i designates increment number.

Heat Balance. It is convenient to base the heat of reaction on the known inlet temperature of the increment.

$$\sum \mathscr{F}_j c_{p_j}(T_{i+1} - T_i) = (-r_{CO})(-\Delta H_{CO})_{T_i} \Delta W$$
$$= (-\Delta \mathscr{F}_{CO})(-\Delta H_{CO})_{T_i} \qquad \text{(CS-5.5)}$$

Algorithm

1. Calculate $(-r_{CO})$ at inlet conditions to increment, i.
2. Calculate $(-r_{CO})_{avg} = (-r_{CO})_i + [(-r_{CO})_i - (-r_{CO})_{i-1}]/2$ (skip for $i = 0$).
3. Calculate new flow rates: $\mathscr{F}_{i+1} = \mathscr{F}_i \pm (-r_{CO})\Delta W$
4. Calculate c_p and $(-\Delta H_{CO})$ @ T_i
5. Calculate ΔT from Eq. CS-5.5.
6. $T_{i+1} = T_i + \Delta T$
7. $y_{i+1} = \mathscr{F}_{i+1}/(\mathscr{F}_T)_i$

An increment of $1°$ can be used and $(-\Delta\mathscr{F}_{CO})$ calculated from Eq. CS-5.5. The mole fraction of each component is calculated and the percentage of CO in the dry gas at the outlet of the increment. An average rate is then calculated based on inlet and outlet conditions of the increment using Eq. CS-5.1. Then ΔW is determined from Eq. CS-5.4. Then proceed to the next increment until the desired outlet CO concentration.

8. Mole fraction CO in dry gas $= [y_{CO}/(1 - y_{H_2O})]_{i+1}$.
9. Go to Step 1.

Note. A ΔW increment of 200 lb was found to never cause a ΔT greater than $1°$. The average time for a case was 1 sec, and thus there was no need to change this increment size. Alternatively, one could select an increment size of $1°$ and calculate ΔX from a heat balance, then ΔW.

Results

The results of the several design cases are summarized in Fig. CS-5.3, where it may be seen that in most cases the optima are not overly sensitive to temperature. It, therefore, becomes an easy task to select a reasonable distribution between the duties of the two reactors by comparing the combinations shown in Table CS-5.1. Outlet temperatures exceeding $790°F$ for the first reactor and $450°F$ for the second by more than $5°$ are excluded.

Design decision obviously depends greatly on catalyst life. If the assumed life of each catalyst is correct, case 2 is slightly preferred over case 1. The choice between various outlet CO concentrations in dry gas for reactor no. 2

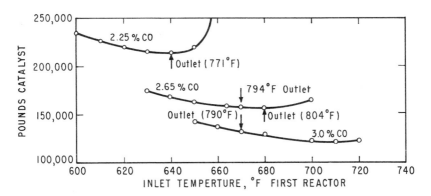

Fig. CS-5.3 Catalyst requirements as function of inlet temperature and outlet CO (first reactor only, $5:1$ steam-to-CO with CO percent at outlet shown).

Table CS-5.1   Results of Calculations

Reactor No.	Outlet CO, %	Temperature, °F In	Temperature, °F Out	Pounds of Catalyst	Cost of Catalyst, $[a]	4-Yr Cost, $
Case 1:						
#1	2.25	640	771	215,800	61,657	
#2	0.3	400	429	150,040	125,033	
Total					186,690	311,725
#1	2.25	640	771	215,800	61,657	
#2	0.35	421	448.3	131,750	109,792	
Total					171,449	281,241
#1	2.25	640	771	251,800	61,657	
#2	0.40	421	447.5	116,900	97.417	
Total					159,074	256,491
Case 2:						
#1	2.65	670	794	157,700	45,057	
#2	0.2	370	400	205,400	171,500	
Total					216,557	388,057
#1	2.65	670	794	157,700	45,057	
#2	0.3	395	429	160,100	133,417	
Total					178,474	311,891
#1	2.65	670	794	157,700	45,057	
#2	0.35	415	448	141,230	117,692	
Total					162,749	280,441
#1	2.65	670	794	157,700	45,057	
#2	0.4	425	457.5	126,140	105,117	
Total					150,174	255,291
Case 3:						
#1	3.0	670	790	132,229	37,780	
#2	0.3	400	439	167,600	139,467	
Total					177,447	317,114
#1	3.0	670	790	132,229	37,780	
#2	0.35	410	448	148,330	123,608	
Total					161,388	284,996
#1	3.0	670	790	132,229	37,780	
#2	0.4	420	457.5	133,000	110,833	
Total					148,613	259,446

[a] Basis: Catalyst Life Reactor No. 1, 4 yr.
    Catalyst Life Reactor No. 2, 2 yr.

remains. As the outlet CO increases more CO must be converted to CH_4 in the methanator, which increases the inerts concentration in the synthesis loop and decreases the ammonia production: For our purposes an approximate rule-of-thumb will be used:

For each 0.1 % increase in inerts there is a 1 % decline in ammonia production.

This fact provides much incentive for maintaining low CO concentrations at the outlet of reactor no. 2. Within these low ranges the decline in production can be offset by modest operating changes at the ammonia converter. An increase in operating pressure can offset the decline in rate caused by the lower reactant partial pressure created by the added inerts. Based on some reasonable steam costs for the plant being considered and turbine water rates, a cost of $16,500/yr for each increase of 0.1 % CO has been estimated. This value will be useful for illustrative purposes, but in a real case actual design calculations on the ammonia converter will yield a more precise value.

Four-year incremental costs of case 2 can be summarized for catalyst and energy with an arbitrary basis of zero for energy at 0.2 % CO and for catalyst at 0.4 % CO.

% CO	Incremental 4-year Costs*
0.2	$132,766
0.3	$122,600
0.35	$124,150
0.40	$132,000

Again, the optimum is flat, and it seems best to select a catalyst charge and outlet CO that will provide good operating flexibility. Since the rate equations are based on used catalyst, early operation will yield low CO values and performance toward the end of the run will produce high CO values. By placing the design on the middle of this region operating flexibility is obtained along with that already provided for raising the temperature.

Decision. 1. Size reactor volume on basis of catalyst required for 0.3 % CO. Table CS-5.2 summarizes the operating conditions.

---

* The results are dramatically altered using 1977 costs ($1.80/MM BTU for energy and $110/cu ft for cat. #2, an increase of 260% for energy and only 47% for catalyst). Operation at 0.2% will now be definitely favored. The incremental 4-year costs are $194,723 for 0.2% CO, $320,701 for 0.3% CO, $393,213 for 0.35% CO, and $475,200 for 0.4% CO.

Table CS-5.2   Summary of Operating Conditions for Reactors No. 1 and No. 2

Reactor no. 1

    Inlet pressure: 26.53 atm (10 psi drop for waste heat boiler and piping)
    Inlet temperature: 670°F
    Outlet temperature: 794 F
    CO in dry gas: 13.0% in, 2.65% out
    Catalyst: 157,700 lb (2253 cu ft)

Reactor no. 2

    Inlet pressure: 25.85 atm (10 psi drop between inlet of #1 and inlet of #2 allowed)[b]
    Inlet temperature: 415°F
    Outlet temperature: 448°F
    CO in dry gas: 2.65% in, 0.30% out
    Catalyst (based on 0.3% CO): 160,100 lb (1789 cu ft)

Compositions (dry gas)

		Mole %	
Component	Inlet #1	Outlet #1 Inlet #2	Outler #2
CO	7.879	1.77	0.24
H_2O	39.394	33.28	31.75
CO_2	4.788	10.90	12.43
H_2	34.424	40.54	42.07
N_2	13.212	13.212	13.212
CH_4	0.303	0.303	0.303
	100.000	100.000	100.000

Average viscosity, cp[a]

In	0.0242		
Out	0.0262		
c_{p_m} (In)	7.98	(Out) 7.87	

Total Flow: 20,460 lb-moles/hr.
Average MW: 15.84.
[a] Based on mixture method in *API Data Book*.
[b] Includes ΔP for boiler feedwater exchanger.

2. Base other calculations and material balance on 0.35% CO so that additional capacity will be provided in all downstream equipment.

Reactor Configuration

Based on reactor inlet conditions (Table CS-5.2) determine mass flow rate that will minimize temperature and concentration gradients between fluid phase and catalyst surface without excessive pressure drop.

Reactor No. 1 @ 670°F and 26.53 Atm

Using Fig. 11.5 and Table CS-5.2

$$N_{Re} = \frac{d_p G}{\mu} = \frac{(1.225)(0.25)G}{(12)(0.0242)(2.42)} = 0.436G$$

$$a_m = \frac{6}{D_p \rho_p} = \frac{(6)(12)}{(0.25)(126)} = 2.286 \text{ ft}^2/\text{lb}$$

$$(-r_{CO}) = \psi k(y_{CO} y_{H_2O} - y_{CO_2} y_{H_2}/K)/379\rho_b$$

$$= (4)\frac{3550}{(379)(70)}\left[(0.79)(0.394) - \frac{(0.048)(0.344)}{19.4}\right]$$

$$= 0.0161$$

$$R_m = \frac{(-r_{CO})M_m}{a_m G} = \frac{(0.0161)(15.84)}{(2.286)G} = \frac{0.112}{G}$$

$$q_m = R_m \frac{(-\Delta H)}{(c_{p_m})} = \frac{(0.112)(9280)}{7.98G} = \frac{130}{G}, °C$$

From Fig. 11.5 it is apparent that G values 500 and above meet the ΔT requirements. There is no need to calculate the Prandtl number. A value of 1.0 is conservative. To avoid calculating a Schmidt number, use a conservative value of 5 for which $\Delta y_i \sim 0.006$ at these mass velocities. This is less than 10% of limiting reactant mole fraction.

Pressure Drop (Reactor No. 1)

From Eqs. 11.7 and 11.8B at average of inlet and outlet,

$$N_{Re} = \frac{D_p G}{\mu} = \frac{(0.25)G}{(12)(0.025)(2.42)} = 0.344G$$

$$f_k = 1.75 + 150\frac{(1-\varepsilon)}{N_{Re}}$$

$$= 1.75 + 150\frac{0.555}{0.344G} = 1.75 + \frac{242}{G}$$

For $G = 500, f_K = 2.23$

$$\frac{\Delta P}{\Delta L} = \frac{f_k G^2}{D_p \rho_f g_c}\left(\frac{1-\varepsilon}{\varepsilon^3}\right)$$

$$\Delta P/\text{ft} = \left[\frac{(2.23)(500)^2}{(0.25/12)(0.483)(32.17)(3600)^2}\right]\left[\frac{0.555}{(0.445)^3}\right]$$

$$= 0.837 \text{ psf/ft or } 5.8 \times 10^{-3} \text{ psi/ft}$$

Can use higher G to reduce equipment size.

Table CS-5.3 L/D Study for Reactor No. 1 (1971 Costs)

$D \times L$, ft[a]	Bed Thickness in.	Bed Height ft	G	ΔP, psi	Vessel Weight, lb	Vessel Cost,[b] $	Inert Support ft^{3c}	Inert Support $	Total $	Savings Vessel $\Delta\$$	Savings Compression $\Delta\$^{d}$	Net Savings $
14 × 19	3.19	14.63	2100	1.26	173,000	114,180	360	3380	117,560	0	0	
13 × 21	2.97	16.97	2440	1.95	156,600	104,874	288	2710	107,584	9,974	(−2,524)	7,450
12 × 24	2.75	19.92	2860	3.12	142,300	98,187	226	2220	100,407	17,153	(−8,077)	9,076
11 × 28	2.53	23.72	3400	5.21	131,400	93,819	174	1640	95,459	22,101	(−17,668)	4,433
10 × 33	2.31	28.70	4130	9.23	121,800	88,914	131	1230	90,144	27,416	(−36,850)	(−9,434)
9 × 40	2.09	35.41	5080	17.13	114,200	84,508	96	900	85,408	32,152	(−75,718)	(−43,566)

[a] Approximately 4 ft added to length in order to account for inert material between tangents at top and bottom and for free space above bed for distributor and work area.

[b] See Appendix B for example calculation. Costs for this type vessel varied from 66 ¢/lb at 14 ft to 74 ¢/lb at 9 ft diameter at time of calculation. ASTM A387D (2¼ Cr-1 Mo) selected for resistance to hydrogen attack (Fig. B.1 Appendix).

[c] Amount of inert support varies with volume of bottom head (¾ in. × ⅞ in. pellets @ $9.40/cu ft).

[d] Compressor delta operating costs are affected lower suction pressure on first-stage of synthesis-gas compressor caused by increased ΔP in bed. Costs are calculated as shown in the Appendix C on the basis of 50 ¢/MM BTU fuel gas and 4 yr payout.

VESSEL SPECIFICATION SHEET					

ITEM NO. _____ R-101
JOB NO. _____
REV. _____ DATE _____
PREPARED BY _____ DATE _____
CHECKED BY _____ DATE _____
APPROVED _____ DATE _____
CUSTOMER _____ PLANT _____

SERVICE OF UNIT FIRST - STAGE SHIFT CONVERTER
NO. REQ'D 1   SIZE: 12'-0" I.D. × 24'-0" TT

CONSTRUCTION

CODE: A S M E Sect. VIII Div. I	STAMP: Yes

PRESS.(PSIG)	MAX OP. 390	NOR OP. 390	DESIGN: 467
TEMP (°F)	MAX OP. 850	NOR OP. 794	DESIGN: 890

WIND VELOCITY:	100	MPH	CORR. ALLOW.: 1/8
STRESS RELIEVE:		X-RAY	Yes
PHYS. & CHEM. TESTS:			

Calculated Thickness : 2.75 in. (Includes corr. all)

MATERIAL	SPEC	MATERIAL	SPEC
SHELL **ASTM**	A 387 D	INTERNALS	
HEADS	A 387 D	DAVIT/HINGE	
SUPPORTS		PLATFORMS	
		LADDER (CAGED)	
		CLIPS	
FLANGES	A 182 Gr F22	PAINTER TROLLEY	
PIPE	A 335 P22	LIFTING LUGS	
COUPLINGS		PAINT	
GASKETS		INSULATION	
BOLTS		INSUL. SUPPORTS	
NUTS			

NOZZLE SCHEDULE

MK.	SERVICE	NO.	SIZE	RATING	FACING	O.S. PROJ
N-1	Manway & Inlet	1	20"	400 #		
N-2	Outlet	1	20"	"		
N-3	Catalyst Discharge	1	12	"		
N-4	Thermowells	4				
N-5						
N-6						
N-7						
N-8						
N-9						
N-10						
N-11						
N-12						
N-13						
N-14						
N-15						
M-1	MANHOLE					

SKIRT OPENINGS:

Note A : Baffles must be removable for catalyst loading

Note B: Outlet Distributor, 18" torus with one continuous 2" slot on bottom side, covered by 6 Mesh SS Screen, 0.047 in. diam. wire.

Fill	Type	cu.ft	pounds
A	3/8"×7/8" Inert Pellets	66.4	5,650
B	Iron Oxide Catalyst	2253	157,710
C	1/2"×5/8" Inert Pellets	66.4	5,780
D	3/4"×7/8" Inert Pellets	288	25,100

REMARKS:
   * Evenly spaced over catalyst bed

Fig. CS-5.4   Process vessel sketch for reactor 1, shift converter.

Reactor L/D Selected and Final Design

The reactor L/D selected above $G = 500$ will depend on minimizing the reactor cost plus incremental compression costs produced by reactor ΔP. A summary for various L/D ratios at constant volume based on the required catalyst is given in Table CS-5.3 for reactor no. 1 and a process vessel sketch is shown in Fig. CS-5.4. The 12-ft diameter reactor was selected as most economical.

Similar calculations for reactor no. 2 were also carried out which resulted in selecting a reactor 12 ft in diameter by 20 ft tangent-to-tangent.

Other Appurtenances

Because of the high sensitivity of the low-temperature catalyst to sulfur poisoning, a guard bed is placed on top of the catalyst bed. The height for good distribution is simply set at 100 times the particle diameter of $\frac{1}{4}$ in., or 2 ft. Guard materials have been reported with long life times (3). These include zinc oxide and special guard solids which also protect against chloride poisoning of the catalyst.

An attractive alternate arrangement for the second converter, though requiring greater capital expenditure, would be two parallel reactors each 8 ft in diameter and 20 ft tangent-to-tangent. Costs could be minimized by mounting one above the other with a common header between them. As the catalyst deactivated one bed could be replaced while the other remained operative without excessive loss of ammonia production. At a wholesale price of $30/ton of NH_3, total shutdown of a plant this size is most costly. The decision must be made on the total economics of the ammonia complex, which is beyond the scope of this particular illustrative case.

REFERENCES

1. H. Bohlbro, *An Investigation of the Kinetics of the Conversion of Carbon Monoxide with Water Vapor*, Gjellerup, Copenhagen, 1966.
2. Girdler Catalysts Technical Data, Chemetron Chemicals Louisville, Kentucky, 1965.
3. J. F. Lombard, *Hydrocarbon Process.*, **48** (8), 111 (1969).
4. O. A. Hougen, K. M. Watson, and R. A. Ragatz, *Chemical Process Principles*, Part II, 2nd ed., Wiley, New York, 1959.

CASE STUDY 106

Ammonia Synthesis

THIS STUDY illustrates the design of multistage, fixed-bed reactors with direct-contact quench between stages. The optimum approach to equilibrium is sought, and convenient graphical procedures that aid in selecting operating conditions are illustrated. This reactor is in the synthesis portion of the total ammonia plant as shown in Fig. CS-5.1 (p. 45).

Problem Statement

Design the converter for an ammonia plant capable of producing 1000 tons/day of liquid ammonia delivered at the battery limits at $-28°F$.

Feed

Synthesis gas from the methanator is delivered to the converter at $100°F$ and 335 psig (see Fig. CS-5.1).

Component	Mole %
H_2	74.03
N_2	24.68
CH_4	0.95
Ar	0.34
	100.00

Catalyst

A triply promoted $(K_2O-CaO-Al_2O_3)$ iron-oxide catalyst will be used. The iron oxide (Fe_2O_3-FeO) is in the form of nonstoichiometric magnetite.

It is made by fusing the magnetite with the promoters. The catalyst is reduced in situ, and the removal of oxygen yields a highly porous structure of iron with promoters present as interphases between the iron crystals and as porous clusters along the pore walls (1). The pores range from $50°A$ to $100°A$, and intraparticle diffusion is thought to occur by the bulk mechanism.

Alumina prevents sintering and corresponding loss of surface area and also bonds the K_2O, preventing its loss during use (1). The K_2O and CaO neutralize the acid character of Al_2O_3. Both K_2O and CaO decrease the electron work function of iron and increase its ability to chemisorb nitrogen by charge transfer to the nitrogen (1).

Properties

Particle Size.   Granules, in size range 6–10 mm.

Bulk Density.   165 lb/cu ft (2.65 kg/liter).

Particle Density.   305 lb/cu ft (4.9 g/cm^3)

Activity loss in service.   30–50% in 3 yr depending on severity of operating conditions and presence of poisons. Catalyst is slowly deactivated at operating temperatures above 985°F (530°C).

Catalyst Poisons.   In addition to poisons shown in Table 2.13, hydrocarbons such as lubricating oils and olefins can crack and plug pores. Sulfur, phosphorous, and arsenic compounds are permanent poisons. Oxygen and listed oxygen compounds should not exceed 15 ppm. Though temporary poisons, they cause crystal growth and attendant area decline. Chlorine compounds form volatile alkalichlorides with promoters (1).

Chemistry and Kinetics

The overall stoichiometric equation is: $\frac{1}{2}N_2 + \frac{3}{2}H_2 \rightarrow NH_3$. Extensive studies of ammonia synthesis on iron catalysts suggest that the reaction occurs through surface imine radicals and the following elementary steps (1,2).

$$N_2(g) \longrightarrow 2N(ads) \tag{1}$$
$$H_2(g) \longrightarrow 2H(ads) \tag{2}$$
$$N(ads) + H(ads) \longrightarrow NH(ads) \tag{3}$$
$$NH(ads) + H(ads) \longrightarrow NH_2(ads) \tag{4}$$
$$NH_2(ads) + H(ads) \longrightarrow NH_3(ads) \tag{5}$$
$$NH_3(ads) \longrightarrow NH_3(g) \tag{6}$$

A rate equation based on nitrogen adsorption as the slow step and the Temkin isotherm is the most commonly used although other forms have been developed that also correlate the data.

Since the effectiveness factor of ammonia catalyst is less than unity in commercial size pellets, it is desirable to develop a rate equation from laboratory data on finely ground catalyst and employ an effectiveness factor correction for other sizes. Ammonia synthesis is another example of an old reaction with sufficient data existent to make this procedure feasible. The following equations in terms of activity have been recommended (3).

Rate on Fine Catalyst

$$2(-r_N) = r_A = 2k\psi \left[K^2 \left(\frac{a_N a_H^{\frac{3}{2}}}{a_A} \right) - \left(\frac{a_A}{a_H^{\frac{3}{2}}} \right) \right], \frac{\text{kg moles NH}_3}{(\text{m}^3\text{cat.})(\text{hr})} \quad \text{(CS-6.1)*}$$

where subscripts A, H, and N refer to ammonia, hydrogen, and nitrogen, respectively, K is the equilibrium constant and ψ is the activity factor to account for effect of particle size on the ultimate surface area after reduction.

The original data are based on reduced particles of 3–6 mm size (4.6 mm effective diameter and 8.6 m²/g). We will use 6–10 mm (5.8 mm effective diameter) for which the area can be estimated from Ref. 15 as 7.5 m²/g. Thus $\psi = 7.5/8.6 = 0.87$.

$$2k = 1.7698 \times 10^{15} \exp\left(-40{,}765/R'T\right) \quad \text{(CS-6.2)}$$

$$\log_{10} K = -2.691122 \log_{10} T - 5.519265 \times 10^{-5} T$$

$$+ 1.848863 \times 10^{-7} T^2 + \frac{2001.6}{T} + 2.6899 \quad \text{(CS-6.3)(5)}$$

where $T = {}^\circ\text{K}$.

Equation CS-6.1 is based on the most complete data on an industrial-type catalyst formulation (4) and predicts ammonia mole fraction with a nominal maximum deviation of 10–20% in the usual pressure range of interest, 150–300 atm. Thus a pressure correction, as suggested on p. 41[1], is included in the value of k as a nominal value.

An effectiveness factor equation developed for the indicated kinetics, rather than pseudo-first-order kinetics, yields a complex expression which

---

* This unusual arrangement results from using the reverse rate constant (ammonia dissociation constant) as the rate constant k.

was solved for a number of values of P and T for a catalyst charge composed of 6–10 mm particles. The values were then summarized in a simple form (3).

$$\eta = b_0 + b_1 T + b_2 X + b_3 T^2 + b_4 X^2 + b_5 T^3 + b_6 X^3 \qquad \text{(CS-6.4)}$$

Values of the constants at 150, 225, and 300 atm are given in Ref. 3.

Since $a_j = f_j/f_j^\circ$ and $f_j^\circ = 1$ atm for gases, $a_j = f_j = y_j v_j P$, where v_j is the fugacity coefficient for component j for which the following equations may be used (6–8) with T in $^\circ$K and P in atm.

$$v_H = \exp\{e^{(-3.8402T^{0.125}+0.541)}P - e^{(-0.1263T^{0.5}-15,980)}P^2$$
$$+ 300[e^{(-0.011901T-5.941)}](e^{-P/300} - 1)\} \qquad \text{(CS-6.5)}$$

$$v_N = 0.93431737 + 0.3101804 \times 10^{-3}T + 0.295896 \times 10^{-3}P$$
$$- 0.2707279 \times 10^{-6} T^2 + 0.4775207 \times 10^{-6}P^2 \qquad \text{(CS-6.6)}$$

$$v_A = 0.1438996 + 0.2028538 \times 10^{-2}T - 0.4487672 \times 10^{-3}P$$
$$- 0.1142945 \times 10^{-5}T^2 + 0.2761216 \times 10^{-6}P^2 \qquad \text{(CS-6.7)}$$

Thus Eq. CS-6.1 for the industrial catalyst becomes in kg moles/(m^{3}cat.) (hr),

$$r_A = \eta\, 1.7698 \times 10^{15} \exp\left(-\frac{40,765}{R'T}\right)\left[K^2 P^{\frac{3}{2}}\left(\frac{v_N y_N v_H^{\frac{3}{2}} y_H^{\frac{3}{2}}}{v_A y_A}\right) - \frac{1}{P}\left(\frac{v_A y_A}{v_H^{\frac{3}{2}} y_H^{\frac{3}{2}}}\right)\right]$$

$$\text{(CS-6.8)}$$

This can be converted to lb moles/(lb cat.) (hr) by multiplying by the factor 3.7836×10^{-4}.

Thermodynamics

For a single reaction, such as ammonia synthesis, the entire message of thermodynamics is conveniently presented on an equilibrium conversion and adiabatic reaction-temperature plot such as shown in Fig. CS-6.1. It is quite clear from this figure that adiabatic beds in series with intermediate cooling will be necessary to attain a conversion in the 15% or over level, as discussed on p. 513[l]. Various strategies can be planned by using an equilibrium approach for this process of 40–50°F; or, more conveniently, rate plots as shown in the next section can be employed.

Direct-contact quench, which is most attractive for a high-pressure system, in contrast to expensive exchangers, will be used. Design strategies can be rapidly developed by graphical constructions using adiabatic lines and quench lines in accordance with the procedure given on p. 530[l]. Alternatively, and perhaps more precisely, such paths may be considered on rate plots, as described on p. 526[l] and shown on p. 69.

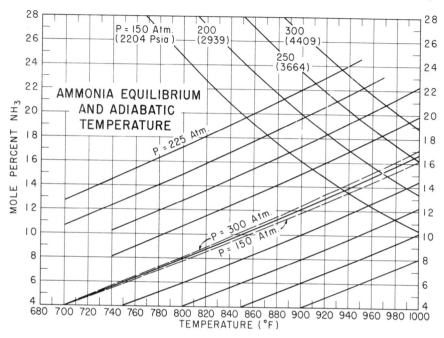

Fig. CS-6.1 Ammonia equilibrium and adiabatic temperature.

Design Equations

Component Balance

Since the feed quantities vary with each bed in a quench reactor, it is convenient to base the component balance on the feed to the first bed.

Mole Balance

Basis: feed to first bed

$$\Delta W \hat{r}_A = \Delta n_{A_F} F_I \qquad \text{(CS-6.9)}$$

or

$$\frac{\Delta W}{F_I} = \frac{\Delta n_{A_F}}{r_A} \qquad \text{(CS-6.10)}$$

$$\Delta n_{A_F} = -\tfrac{3}{2}\Delta n_{H_F} = -\tfrac{1}{2}\Delta n_{N_F} \qquad \text{(CS-6.11)}$$

where F_1 is the feed to first bed and n_{A_F} is the moles of ammonia per mole of feed to first bed.

As quench is added between beds, the value of n for each component changes depending on the amount of quench.

Basis: per mole of feed to first bed

$$(n_{A_F})_e^I = (n_{A_F})_0^I + \Delta n_{A_F}$$

$$(n_{A_F})_0^{II} = (n_{A_F})_e^I + \gamma_{II}(n_{A_F})_0^I \qquad \text{(CS-6.12)}$$

where γ_{II} is the fraction of feed to first bed for quench between first and second bed, $(n_{A_l})_0$ is the inlet ammonia per mole of feed to first bed for numbered bed, and $(n_{A_l})_e$ is the outlet ammonia per mole of feed to first bed from numbered bed. For nth bed

$$(n_{A_F})_0^n = (n_{A_F})_e^{n-1} + \gamma_{n-1}(n_{A_F})_0^{n-1} \qquad \text{(CS-6.13)}$$

Similar equations for other components based on Eq. CS-6.11 are apparent.

Heat Balance (between i and i + 1 positions)

Basis: per mole of feed to first bed

$$\left(\sum n_{j_F} H_{f_j}{}^\circ\right)_i + \sum \left[(n_{j_F})_i \int_{298}^{T_i} c_{p_j} dT\right]$$

$$= \left(\sum n_{j_F} H_{f_j}{}^\circ\right)_{i+1} + \sum \left[(n_{j_F})_{i+1} \int_{298}^{T_{i+1}} c_{p_j} dT\right] \qquad \text{(CS-6.14)}$$

or in terms of heat of reaction

$$\sum \left[(n_{j_F})_{i+1} \int_{T_i}^{T_{i+1}} c_{p_j} dT\right] = \Delta n_{A_F}(-\Delta H_A)_{T_i, P} \qquad \text{(CS-6.15)}$$

It is sufficient to consider the pure component enthalpies of the various gases additive. In this system ideal-gas heat capacity equations apply to all but ammonia, for which the following was used in cgs units (11).

$$c_{p_A} = 6.5846 - 0.61251 \times 10^{-2}T + 0.23663 \times 10^{-5}T^2$$
$$- 1.5981 \times 10^{-9}T^3 + 96.1678 - 0.067571P + (-0.2225$$
$$+ 1.6847 \times 10^{-4}P)T + (1.289 \times 10^{-4} - 1.0095 \times 10^{-7}P)T^2$$

where $T = {}^\circ K$ and P is in atm.

Other c_p equations are the same as used for Case Study 105. The expression for the heat of reaction based on this equation and the other ideal heat capacities is (11):

$$(\Delta H_\text{A})_{T,P} = -9184.0 - 7.2949T + 0.34996 \times 10^{-2}T^2 + 0.03356$$
$$\times 10^{-5}T^3 - 0.11625 \times 10^{-9}T^4 - (6329.3 - 3.1619P)$$
$$+ (14.3595 + 4.4552 \times 10^{-3}P)T - T^2(8.3395 \times 10^{-3}$$
$$+ 1.928 \times 10^{-6}P) - 51.21 + 0.14215P, \text{cal/g mole NH}_3$$

$$(\text{CS-6.16})$$

Establishing Purge, Recycle, and Total Feed

Referring to Fig. CS-6.2, the following mole balances may be formulated.

Inert Balance

$$(y_\text{I})_\text{F} F_\text{F} = (y_\text{I})_\text{PU} F_\text{PU} \qquad (\text{CS-6.17})$$

$N_2 + H_2$ *Balance*

$$(y_{\text{N}_2+\text{H}_2})_\text{F} F_\text{F} = (y_{\text{N}_2+\text{H}_2} + 2y_\text{A})_\text{PU} F_\text{PU} + 2F_\text{P} \qquad (\text{CS-6.18})$$

$$[(y)_{\text{N}_2+\text{H}_2} + y_\text{I} + y_\text{A}]_\text{PU} = 1 \qquad (\text{CS-6.19})$$

where F refers to fresh feed, P to product, and PU to purge; y_I is mole fraction inerts.

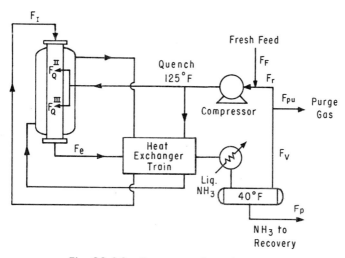

Fig. CS-6.2 Converter and recycle system.

Based on phase equilibrium at the separator,

$$(f_A)_{LIQ} = (v_A)_{LIQ}(P_A)_{VP} = (f_A)_g \qquad \text{(CS-6.20)}$$

where $(P_A)_{VP}$ is the vapor pressure of ammonia at temperature and pressure of the system and $(f_A)_g$ is the ammonia fugacity in the vapor phase.

The fugacity of ammonia in the vapor phase can be calculated using the Redlich–Kwong equation of state (9,10).

$$\ln \frac{(f_A)_g}{P y_A} = (z_m - 1)\frac{B_A}{B} - \ln(z_m - BP)$$

$$- \frac{A^2}{B}\left(\frac{2A_A}{A} - \frac{B_A}{B}\right)\ln\left(1 + \frac{BP}{Z}\right) \qquad \text{(CS-6.21)}$$

where $A = \sum A_j y_j$, $B = \sum B_j y_j$, $A_j = 0.6541/T_{r_j}^{1.25} P_{cr_j}^{0.5}$, and $B_j = 0.0867/T_{r_j} P_{cr_j}$.

$$z_m = \frac{1}{1-h} - \frac{A^2}{B}\frac{h}{1+h} \qquad \text{(CS-6.22)}$$

where $h = BP/z_m$, z_m is the compressibility factor, P_{cr} is the critical pressure, and T_r is the reduced temperature.

This requires trial-and-error since z_m and f_A depend on composition. At a set inerts composition and H_2/N_2 ratio, an ammonia mole fraction in the vapor phase is assumed. Then Eqs. CS-6.20 and CS-6.21 are applied with acceptable convergence defined as

$$\frac{(f_A)_g - (f_A)_{LIQ}}{(f_A)_g} \lesssim 10^{-3}$$

Design Studies—Rapid Overview

The large number of interrelated variables involving inerts, recycle, separator, and reactor conditions and number of beds makes it necessary to limit or set some of these consistent with good practice. Any attempt to search the entire universe of values is not rational particularly when low sensitivity of results to many changes and accuracy of the rate data are considered. Because composition of inerts affects reactor performance, recycle, and separator equilibrium, it is convenient to base cases for study on inerts composition. The inerts composition of the total feed to the converter (fresh feed plus recycle) can be specified at various values in a range known to be consistent with good practice (8–15%). One might, for example, select an intermediate value of 12% for the purpose of studying various operating pressures. Product recovery and quench temperatures can also be set as well as ammonia production rate, number of beds, and inlet and outlet temperature of each bed. These latter temperatures are most conveniently selected by generating a rate plot, auch as presented in Fig. CS-6.3. Graphical constructions are made using the adiabatic line and quench slopes, as discussed on p. 530[1]. Since the feed and quench composition depend somewhat on recycle rates, the first approximations must be made on the basis of a guess for this value (3–4%). The first run of the computer program on a case will establish not only this composition but also the slopes of the quench and adiabatic lines with greater precision so that other cases of bed temperatures may be considered. It will become apparent that catalyst is minimized by avoiding too close an approach to equilibrium and by selecting as the inlet temperature of the next bed a point along the quench line where the adiabatic temperature line just becomes parallel with a constant rate line. Usually this occurs over a range. In graphically determining bed temperature strategies

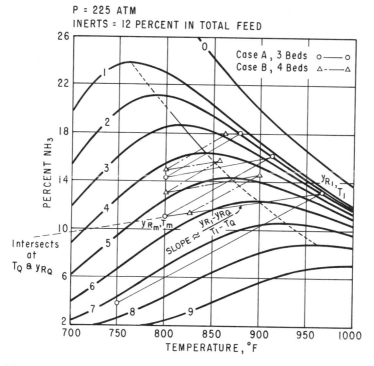

Fig. CS-6.3 Graphical solutions on computer-generated rate plot. ($r \times 10^4 = 50$ lb-moles/(lb cat.) (hr) for contour **1**, 71 for **2**, 100 for **3**, 141 for **4**, 200 for **5**, 282 for **6**, 398 for **7**, 562 for **8**, and 795 for **9**. These rates for 3–6 mm catalyst. Multipy by 0.87 for 6–10 mm catalyst.)

for trial, it is most convenient at a given pressure to select a reasonable low rate as the closest approach (e.g., line 2 on Fig. CS-6.3 for 225 atm) and a corresponding outlet percentage NH_3 at a reasonable outlet temperature (e.g., 18% at 225 atm and 875°F). In this manner the reaction paths straddle the locus of maximum rates. Various cases in these general regions may then be calculated and compared. Referring to Fig. CS-6.3, a graphical procedure involving beginning at the set outlet conversion is followed, and one proceeds backward through the reactor. The adiabatic line A is followed until the rate begins to decrease and then the quench line is drawn and the process repeated. Two such cases are sketched. The inlet and outlet temperatures approximated from these constructions are then used to provide cases for computer calculations.

Alternatively, a computer routine can be devised to search for the minimum catalyst requirement by changing bed inlet and outlet temperatures systematically. The graphical selection of reasonable cases, however, is so rapid and efficient that the effort in developing this additional subroutine may not be warranted.

Design Algorithm

Input data are: number of beds, pressure at reactor inlet (P) and at separator ($\approx 0.92P$), product–recovery separator temperature, ammonia product rate, percent inerts in converter inlet, temperature of quench, and inlet and outlet temperature of each bed.

1. Select and set as independent variable the percentage inerts in purge.
2. At set inerts in purge determine y_{H_2}, y_{N_2}, and y_A from equilibrium at separator.
3. Determine purge flow rate and fresh feed using Eqs. CS-6.17–CS-6.19 to solve for F_{PU} and F_F.
4. Determine recycle (F_r) from inerts balance on reactor total feed (F_T), fresh feed (F_F) and recycle

$$(y_I)_T F_T = (y_I)_F F_F + (y_I)_r F_r \qquad \text{(CS-6.23)}$$

$$(y_I)_r = (y_I)_{PU} \qquad \text{(CS-6.24)}$$

$$F_T = F_F + F_r \qquad \text{(CS-6.25)}$$

where y is the mole fraction inerts. Sub F indicates fresh feed.
5. Calculate first bed at point i in reactor.
   (a) Calculate rate, $\hat{r}_{A_i} = f(P, T_i, y_i)$ (pressure drop will be minimized and average P for bed is used).
   (b) Calculate average rate for increment by linear extrapolation

$$\bar{\hat{r}}_{A_i} = \hat{r}_{A_i} + \tfrac{1}{2}(\hat{r}_{A_i} - \hat{r}_{A_{i-1}}) \qquad \text{(CS-6.26)}$$

(c) Calculate Δn_{A_F}, Δn_{H_F} and Δn_{N_F} for increment $\Delta W/F_I$ using Eqs. CS-6.9–CS-6.11, where F_I is feed to first bed.

(d) Calculate $(-\Delta H)_A$ at P, T_i (Eq. CS-6.16 and $C_{p_j} = f(T_i, P)$.

(e) Calculate T_{i+1} from heat balance over increment using Eq. CS-6.15.

(f) Go to step 5(a).

6. Calculate fraction quench required to produce temperature at inlet to bed no. II (T_0^{II})

Basis: T_0^{II} and 1 mole feed to first bed

$$\sum [(n_j)_e^I c_{p_j}(T_e^I - T_0^{II})] + \sum_i [\gamma_{II}(n_j)_0^I c_{p_j}(T_Q^{II} - T_0^{II})] = 0$$

where T_Q is the quench temperature and n_j is the moles of component j per mole of feed to first bed.

$$\gamma_{II} = \frac{\sum [(n_j)_e^I c_{p_j}(T_e^I - T_0^{II})]}{\sum [(n_j)_0^I c_{p_j}(T_0^{II} - T_Q^{II})]} \qquad \text{(CS-6.27)}$$

7. Calculate composition and mole fractions of inlet to bed II

$$(n_j)_0^{II} = (n_j)_e^I + \gamma_{II}(n_j)_0^I$$

$$y_j^{II} = \frac{(n_j)_0^{II}}{\sum (n_j)_0^{II}}$$

8. Proceed as in item 5 through bed II, etc.

9. At outlet of last bed calculate ratio F_I/F_T where F_T is the total feed.

$$F_T = F_I(1 + \gamma_{II} + \gamma_{III} + \cdots + \gamma_n)$$

$$\frac{F_I}{F_T} = \frac{1}{1 + \gamma_{II} + \gamma_{III} + \cdots + \gamma_n} \qquad \text{(CS-6.28)}$$

where γ is the fraction of feed to first bed as quench at entrance to indicated bed.

10. Calculate flow rates F_I, F_Q^{II}, F_Q^{III}, etc. F_e, and F_v where F_e is the outlet flow in moles/time from reactor. F_T is known from Item 4. From CS-6.28

$$F_I = \left(\frac{1}{1 + \gamma_{II} + \gamma_{III} + \cdots + \gamma_n}\right)F_T \qquad \text{(CS-6.29)}$$

$$F_e = \sum (n_j)_e F \qquad \text{(CS-6.30)}$$

$$F_Q^{II} = \gamma_{II}F_I, \ F_Q^{III} = \gamma_{III}F_I, \text{ etc.} \qquad \text{(CS-6.31)}$$

$$F_v = F_e - F_P \qquad \text{(CS-6.32)}$$

11. Calculate vapor composition from separator for all components except NH_3

$$F_v(y_j)_v = (y_j)_e F_e P \qquad \text{(CS-6.33)}$$

$$(y_A)_v = 1 - (y_N + y_H + y_A + y_M)_v \qquad \text{(CS-6.34)}$$

where y_M is the mole fraction of methane.

12. Compare with vapor compositions originally calculated. Seek the minimum of

$$f(x) = \sum [y_{j_v} F_v - y_{j_r}(F_{PU} + F_r)]^2 \qquad \text{(CS-6.35)}$$

13. Calculate function given by Eq. CS-6.35 and select new value of percent inerts in purge and return to item 1. The selection of a new value is guided by a unidimensional search subroutine.

14. Convergence Criterion

$$\frac{(y_I)_{new} - (y_I)_{old}}{(y_I)_{old}} < 10^{-2} \qquad \text{(CS-6.36)}$$

15. Calculate W for each bed

$$W_n = \left(\frac{W}{F_I}\right)_n F_I \qquad \text{(CS-6.37)}$$

This describes one case as set by input variables.

Design Cases

Three pressures (150, 225, and 300 atm) will be studied at 12% inerts in the total feed and 40°F separator temperature. For each the catalyst will be minimized and a maximum allowable temperature of 970°F used as a safe distance from the maximum catalyst-use temperature. Since the rate data will be considered to represent lined-out activity after the early high activity, the amounts of catalyst determined for each case will be increased by the factor 1/0.70 to allow for full production at end-of-run conditions. The cost differentials between the several pressures will be determined and the final design pressure selected.

With converter bed sizes and number fixed, an operating study at various temperatures, purge rates, inert contents, and catalyst activity factor can be conducted. A program in which catalyst quantities are set and operating conditions varied is used. In this manner the affect of operating variables on product rate can be determined, and the flexibility of the proposed design assessed.

Required Production

Basis: 1000 tons/day

$$\frac{(1000)(2000)}{(24)(17)} = 4902 \text{ lb moles/hr}$$

Assume 99% recovery

$$\frac{4902}{0.99} = 4952 \text{ lb moles/hr}$$

Recycle Loop ΔP and Separator Pressure

The recycle is handled in one wheel and modern compressors can be designed for a head of 15,000 ft lb_f/lb_m per stage (wheel). It is possible, however, to design the stage for the recycle to operate at a minimum of 50% of this value. The values given below are based on this assumption, and the corresponding separator pressures are calculated using a 1% ΔP between the separator and the compressor.

Pressure, atm.	Loop ΔP, atm.	Separator Press., atm.	Nominal value, atm.
150	13.8	137.5	138
225	20	207.3	207
300	25.4	277.6	278

Results

The results for the cases selected for the three pressures are given in Tables CS-6.1–CS-6.3, and temperatures in the associated heat exchanger train were determined by heat and material balances.

Vessel Diameter and Bed Lengths

Calculate lengths and diameter on the basis of catalyst amounts shown in Tables CS-6.1–CS-6.3 increased by 1/0.7 to satisfy end of run conditions. For this study a bed $\Delta P \approx 0.02P$ will be used. Other values could be studied in a similar manner. Pressure drop in the compressor loop, of which the converter is a part, sets a portion of the power requirements and various ΔP's must be considered. Higher values reduce vessel and catalyst costs but increase power consumption. Excessively low values cause maldistribution and temperature gradients between the catalyst and bulk fluid.

Table CS-6.1 Ammonia Synthesis Converter[a] (150 atm)

		Material Balance									
		Hydrogen		Nitrogen		Ammonia		Methane		Argon	
Stream	Total Flow	Frac.	Flow	Frac.	Flow	Frac.	Flow	Frac.	Flow	Frac.	Flow
Fresh feed	10948.7	0.740300	8105.3	0.246800	2702.1	0.000000	0.0	0.009500	104.0	0.003400	37.2
Recycle	49382.0	0.594743	29369.6	0.198248	9789.9	0.063264	3124.1	0.105859	5227.5	0.037886	1870.9
In	60330.7	0.621158	37474.9	0.207059	12492.0	0.051783	3124.1	0.088372	5331.5	0.031628	1908.1
Outlet	55311.3	0.541405	29945.9	0.180475	9982.3	0.147230	8143.5	0.096392	5331.5	0.034498	1908.1
Sep. Vapor	50359.3	0.594644	29945.9	0.198222	9982.3	0.063374	3191.5	0.105870	5331.5	0.037890	1908.1
Purge	982.6	0.594743	584.4	0.198248	194.8	0.063264	62.2	0.105859	104.0	0.037886	37.2

Reactor Summary

Quench Temperature = 125.00°F

Bed	Lb Cat.	Temp. In	Temp Out	Mole Fraction		Frac. Feed	Total Feed into Bed	Flow out of Bed	Quench at Bed Outlet
				NH$_3$ In	NH$_3$ Out				
1	110481.5	812.00	942.02	0.051783	0.103940	0.735448	44370.1	42273.8	9125.9
2	179476.9	800.00	888.04	0.094680	0.131748	0.151265	51399.7	49716.2	6834.7
3	190791.3	800.00	858.01	0.122083	0.147230	0.113287	56550.9	55311.3	

Total Catalyst = 480749.7 lb

[a] Reactor pressure = 150.00 atm.
Fraction inerts in reactor inlet = 0.12.
Ammonia recovered at 138.000 atm and 40.00°F at a rate of 4952.0 lb moles/hr.
Flows given in lb moles/hr and temperature in °F.
Design catalyst loadings are $\frac{1}{0.70}$ (total shown).

Table CS-6.2 Ammonia Synthesis Converter[a] (225 atm)

Material Balance

Stream	Total Flow	Hydrogen Frac.	Flow	Nitrogen Frac.	Flow	Ammonia Frac.	Flow	Methane Frac.	Flow	Argon Frac.	Flow
Fresh feed	10836.9	0.740300	8022.6	0.246800	2674.6	0.000000	0.0	0.009500	103.0	0.003400	36.8
Recycle	30852.1	0.592868	18291.2	0.197623	6097.1	0.051891	1600.9	0.116076	3581.1	0.041543	1281.7
In	41689.0	0.631192	26313.8	0.210406	8771.6	0.038402	1600.9	0.088372	3684.1	0.031628	1318.5
Outlet	36691.6	0.512859	18817.6	0.170963	6272.9	0.179834	6598.4	0.100409	3684.1	0.035036	1318.5
Sep. Vapor	31739.6	0.592875	18817.6	0.197636	6272.9	0.051872	1646.4	0.116074	3684.1	0.041542	1318.5
Purge	886.9	0.592868	525.8	0.197623	175.3	0.051891	46.0	0.116076	103.0	0.041543	36.8

Quench Temperature = 125.00°F

Reactor Summary

Bed	Lb Cat.	Temp. In	Temp Out	Mole Fraction NH₃ In	NH₃ Out	Frac. Feed	Total Feed into Bed	Flow out of Bed	Quench at Bed Outlet
1	70887.1	750.00	970.07	0.038402	0.130499	0.684257	28526.0	26202.1	7178.6
2	89001.2	800.00	913.01	0.110693	0.161874	0.172195	33380.7	31910.3	5984.4
3	101695.2	800.00	880.02	0.142375	0.179834	0.143549	37894.7	36691.6	

Total Catalyst = 261583.5 lb

[a] Reactor pressure = 225.00 atm.
Fraction inerts in reactor inlet = 0.12.
Ammonia recovered at 207.00 atm and 40.00°F at a rate of 4952.0 lb moles/hr.
Flows given in lb moles/hr and temperature in °F.
Design catalyst loadings are $\frac{1}{6.76}$ (total shown).

75

Table CS-6.3 Ammonia Synthesis Converter[a] (300 atm)

Material Balance

Stream	Total Flow	Hydrogen Frac.	Hydrogen Flow	Nitrogen Frac.	Nitrogen Flow	Ammonia Frac.	Ammonia Flow	Methane Frac.	Methane Flow	Argon Frac.	Argon Flow
Fresh feed	10743.5	0.740300	7953.4	0.246800	2651.5	0.000000	0.0	0.009500	102.1	0.003400	36.5
Recycle	21728.1	0.584507	12700.2	0.194836	4233.4	0.047701	1036.5	0.127371	2767.5	0.045585	990.5
In	32471.6	0.636053	20653.7	0.212029	6884.9	0.031919	1036.5	0.088372	2869.6	0.031628	1027.0
Outlet	27481.4	0.479171	13168.3	0.159737	4389.8	0.219302	6026.7	0.104419	2869.6	0.037371	1027.0
Sep. Vapor	22529.4	0.584493	13168.3	0.194847	4389.8	0.047704	1074.7	0.127371	2869.6	0.045585	1027.0
Purge	801.3	0.584507	468.4	0.194836	156.1	0.047701	38.2	0.127371	102.1	0.045585	36.5

Reactor Summary

Quench Temperature = 125.00°F.

Bed	Lb Cat.	Temp. In	Temp. Out	Mole Fraction NH$_3$ In	NH$_3$ Out	Frac. Feed	Total Feed into Bed	Flow out of Bed	Quench at Bed Outlet
1	41409.0	700.00	970.08	0.031919	0.149635	0.580973	18865.1	16933.4	
2	72347.9	750.00	915.02	0.116166	0.195490	0.207172	23660.6	22090.7	6727.2
3	103286.8	750.00	875.02	0.156648	0.219302	0.211854	28969.9	27481.4	6879.2

Total Catalyst = 217043.7 lb

[a] Reactor pressure = 300.00 atm.
Fraction inerts in reactor inlet = 0.12
Ammonia recovered at 278.00 atm and 40.00°F at a rate 4952.0 lb moles/hr.
Flows given in lb moles/hr and temperature in F.
Design catalyst loadings are $\frac{1}{0.70}$ (total shown).

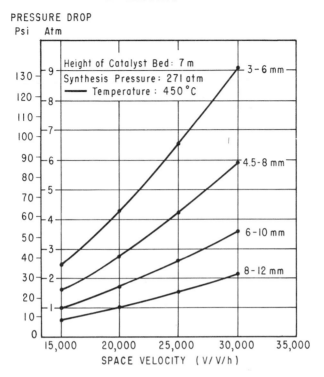

PRESSURE DROP

Fig. CS-6.4 Pressure-drop data for granular ammonia catalyst (multiply by 1.9 to obtain ΔP for dense-packed arrangement for design). [For other conditions:

$$\frac{\Delta P_2}{\Delta P_1} = \left(\frac{Z_2}{Z_1}\right)^{2.85} \left[\frac{(SV)_2}{(SV)_1}\right]^{1.85} \left(\frac{P_1}{P_2}\right)\left(\frac{T_2}{T_1}\right)\left(\frac{M_2}{M_1}\right)^{0.85}$$

(SV) is the space velocity $V/V/h$ measured at 0°C and 1 atm, M_2 and M_1 are the molecular weights of mixture]. Reproduced by permission: Haldor Topsoe A/S, Copenhagen, Denmark, A. Nielsen, *An Investigation on Promoted Iron Catalysts for the Synthesis of Ammonia*, 3rd ed., Gjellerups, Copenhagen, 1968.

Set the diameter using bed *II* as representing an average condition. Pressure-drop data obtained for the particular granular catalyst are given in Fig. CS-6.4.

Example Case 2: (avg. flow = 32646 lb moles/hr, ρ_b = 165, T = 856°F (731°K), P = 225 atm

$$(SV)_2 = \frac{(32640)(359)}{89001/(0.7)(165)} = 15206 \; V/V/\text{hr}$$

From Fig. CS-6.4 @ (SV) = 15206 and catalyst 6–10 mm

$$\Delta P_1 = (1.9)(1.02) = 1.94 \text{ atm}$$

Let

$$\Delta P_2 = (225)(0.02)\left(\frac{89001}{261584}\right) = 1.53 \text{ atm}$$

$$\frac{1.53}{1.94} = \left[\frac{Z_2}{(7)(3.28)}\right]^{2.85} \left(\frac{271}{225}\right)\left(\frac{731}{723}\right)$$

$$Z_2 = 19.72 \text{ ft}$$

$$D = \left[\frac{(89001)(4)}{(0.7)(165)(\pi)(19.72)}\right]^{\frac{1}{2}} = 7.05 \text{ ft}$$

ΔP and lengths for beds *I* and *III* are 15.7 ft and 0.9 atm and 22.56 ft and 2.2 atm, respectively.

Check ΔT between bulk fluid and catalyst to see if mass velocity is adequate. The pressure drop data in Fig. CS-6.4 was plotted as $\Delta P/Lu_s$ versus G.

$G = \dfrac{(SV)(M_m)}{2.24 \times 10^4} L$ g/cm^2 sec	$u_s = \dfrac{G}{\rho}$ cm/sec	$\dfrac{\Delta P}{Lu_s}, \dfrac{\text{dynes sec}}{\text{cm}^4}$
1.125	28.5	50.8
1.50	38.0	66.6
1.88	47.6	82.2
2.25	57.0	94.0

A plot of $\Delta P/(Lu_s)$ versus G yields the slope b in the expression, $\Delta P/(Lu_s) = a + bG$, as described on p. 113[1].

$$b = 37.7, \varepsilon = 1 - \frac{\rho_b}{\rho_p} = 1 - \frac{165}{305} = 0.459$$

$$a_s = \frac{b\varepsilon^3}{(\beta/8)(1 - \varepsilon)} = \frac{(37.7)(0.179)}{0.48} = 14.1 \text{ cm}^{-1}$$

$$a_m = a_s/\rho_p = (14.1)(2.54)(12)/305 = 1.41 \text{ ft}^2/\text{lb}$$

$$D_p = \frac{6}{0.41} = 0.43 \text{ cm or } 4.3 \text{ mm}$$

$$d_p \approx (1.35)(0.43) = 0.58 \text{ cm or } 0.019 \text{ ft}$$

Mixture Properties at inlet to case 2, bed I:

$$\mu = 0.026 \text{ cp} \quad (API \ Data \ Book, \text{Proc. 71 B2.1})$$
$$c_p = 7.7 \text{ BTU/lb mole }°F, \text{ M} = 10.57 \quad (14)$$
$$\lambda_f = 0.132 \text{ BTU/hr ft}°F \quad (API \ Data \ Book, \text{Proc. 12B2.1})$$

Rate at inlet from Fig. CS-6.3 is 0.0394 lb mole/(lb cat.) (hr):

$$\left(\frac{c_p\mu}{\lambda_f}\right)^{\frac{2}{3}} = \left[\frac{(7.7)(0.026)(2.42)}{(10.57)(0.132)}\right]^{\frac{2}{3}} = 0.494$$

From Fig. 11.5

$$R_m = \frac{(-r_N)M_m}{a_m G} = \frac{(0.0394)}{(1.41)(28526/39)} = 3.82 \times 10^{-5}$$

$$q_m = \frac{R_m(-\Delta H)}{c_{p_m}} = \frac{(3.82 \times 10^{-5})(22000)(1.8)}{7.7} = 0.197$$

$$N_{Re} = \frac{0.019[(28526)(10.57)/39]}{(0.026)(2.42)} = 2335$$

$$\Delta T = 4\left(\frac{0.46}{0.4}\right)(0.5) = 2.3°F \text{ within accuracy of data.}$$

$\therefore$ G adequate and diameter selected OK.

Vessel Design

High pressure vessels require the attention of experts in design and fabrication. Ammonia converters in particular, with the many internals and high pressure, present complex mechanical design problems that are wisely assigned to firms specializing in these units. Our focus in this study will be confined only to estimating a vessel weight so that approximate vessel cost can be determined based on data from fabricators on previously built converters of similar size. To reduce vessel thickness a cooling jacket will be used through which will be passed cool synthesis gas that has been only partially exchanged in the heat recovery train. The temperature rise will be about 25°F and the corresponding heat removed only 3 % of that generated in the reactor. The adiabatic assumption in the original calculations remains justifiable although it is possible to add a loss term to the model if accuracy of rate data being used should justify such action. The style vessel will be similar to that shown in Fig. 11.10, and cost will be based on 70¢/lb of vessel including internals and nozzles (1972 costs). The inner shell will be constructed of stainless steel (SA 240 type 316) and the outer shell of carbon steel (SA 204 grade A).

Since the cooling stream (Fig. CS-6.2) is at a higher pressure than the reacting gases inside the shell, this thickness must be calculated using Fig. UHA 28.2 of the ASME Code for vessels under external pressure. Final design would include a shell of increasing thickness toward the bottom, but for estimating purpose we will determine a single thickness based on maximum design pressure.

The results of calculations on both inner and outer shells are summarized in Table CS-6.4, together with estimated costs. Refer to Appendix B and the ASME Code for design procedures.

Compressor Costing and Selection

The crucial nature and high cost of the synthesis and recycle compressor chain demands serious study by a team of experts including compressor manufacturer's representatives. The latter can often suggest even modest changes in design conditions which are perfectly acceptable that will fit better into a given compressor design and save much capital and operating cost.

Higher pressure systems require three barrel casings (frames) on a single drive shaft, and lower pressure, two. Obviously, the fewer casings the lower the capital, and maintenance cost. The trend in compressor development is toward more horsepower per wheel. For purposes of cost estimating, we will employ 2 casings for 150 atm and 3 casings for 225 atm and 300 atm, realizing that it may be possible to include the 225 atm in two casings as well and reduce its cost.

The results of these calculations, which were made in accordance with the procedures outlined in Appendix C, are given in Table CS-6.5.

Higher pressures reduce the refrigeration requirements for ammonia recovery; and refrigeration compression costs, as summarized in Table CS-6.6, must also be included.

Selecting Operating Pressure

The major equipment and operating costs for which significant cost differences exist are summarized in Table CS-6.7. A 3-yr basis or payout was selected which corresponds to the assumed life of the catalyst.

As has been shown previously on a slightly different basis (13), there is little difference between the three cases. The 150 atm and 225 atm cases are essentially the same, but the higher energy requirements for the 225 atm case, provide ominous warnings in an era of constantly rising fuel costs. It seems preferable, therefore, to select the lower pressure case (150 atm), for

Table CS-6.4 Converter Summary (1972 Costs)

Case	Nominal Pressure atm	ID Inner Shell, ft	Total Bed Height, ft	Total Bed ΔP, atm	Inner Shell[a]		Outer Shell[b]			Total Weight lb	Vessel Cost $[c]
					Height ft	Weight lb	ID ft	Height ft	Weight lb		
1	150	10.4	49	3	55	135,147	10.7	56	944,307	1.08×10^6	756,000
2	225	7.05	58	4.6	64	100,958	7.45	65	794,448	0.895×10^6	626,500
3	300	5.6	76.3	6.1	82	105,799	6.1	83	924,911	1.031×10^6	721,700

[a] Design for 1020°F and 2 (bed ΔP) to allow for partial plugging. Use Fig. UHA 28.2 in ASME Code for vessels with exterior pressure. Use ASTM A240 Type 316. See Appendix B for costing technique.

[b] Design for 600°F and 1.1 (nominal pressure). Use a 0.5 Mo steel (see Fig. B.1 in Appendix). ASTM A204 grade A (avoid prolonged exposure above 875°F. See Appendix B for discussion on vessel costing.

[c] $0.7 lb of vessel weight without nozzles and other internals.

Table CS-6.5   Synthesis and Recycle Compressor Summary (1972 Costs)

Nominal Pressure Atm	No. Comp. Casings	Total HP	Compressor Cost, $	Turbine Cost, $	Total $	Energy Consumption Equivalent Fuel, BTU/hr
150	2	17,370	339,500	403,700	743,200	9.71×10^7
225	3	21,460	428,500	446,000	874,500	12.10×10^7
300	3	22,240	443,800	452,400	896,200	12.50×10^7

Notes:
1. Fresh feed compressed from 365 psia and 100°F to 945 psia in first casing, 150-atm alternate, and 888 psia and 853 psia for 225-atm and 300-atm alternates, respectively.
2. Recycle enters last casing at 40°F and 2120, 3180, and 4230 psia, respectively for the three alternates.
3. Discharge pressure of last casings are 2260, 3380, and 4500 psia, respectively.
4. Intercooling is at 40°F between casings.
5. Isentropic work was determined using an HTS diagram prepared from literature data for $3:1$ H_2-N_2 (12) that were found satisfactory and most convenient for this purpose. Actual horsepower was determined based on polytropic efficiency of 74% corrected to an isentropic efficiency, as described in Appendix C.
6. Steam turbine estimates are based on dual-turbine drives, 70% of power in 1500 psig steam @ 900°F to 600 psig and 30% in 600 psig @ 695°F to 4 in. Hg absolute. Efficiency assumed to be 75%.
7. Energy consumption based on 85% efficiency at boiler plant. No credit for excess 600 psig steam made in first turbine.

it will become even more attractive as fuel costs inevitably rise. Obviously, longer catalyst life and higher activity should be important goals in this business. Higher conversions reduce power requirements by lowering the amount of recycle. A catalyst active at a lower temperature would be most attractive for this exothermic reaction.

Earlier studies (13) indicated that higher pressures are required for plants operating at 1500 tons per day capacity. This was necessitated by the impractical converter sizes encountered for this larger capacity at 150 atm. Again, however, escalating fuel costs suggest reevaluation of this conclusion. It is possible that the new horizontal converters (see p. 518[1]) could be specified at lower pressures without encountering impractical configurations.

Although the operating pressure was based on a particular inert level, quench temperature, and approach to equilibrium, other conditions do not substantially alter the conclusions. The next step in design would be a study of other operating conditions at the selected pressure.

Table CS-6.6   Refrigeration Compression Costs (1972 Costs)

Nominal Converter Pressure, Atm	Total Refrigeration HP	Compressor Cost	Turbine Cost	Total $	Energy Consumption Equiv. Fuel, BTU/hr
150	7300	209,200	152,500	361,700	5.05×10^7
225	5199	191,200	118,900	310,100	3.96×10^7
300	4154	184,200	102,100	286,300	3.17×10^7

By lowering the separator temperature, for example, the inlet NH_3 in the recycle will be reduced. The outlet concentration of ammonia can be reduced accordingly while maintaining the same conversion level. The net result will be a smaller catalyst charge because the average rate will be higher. There may be merit in limiting the maximum operating temperature and thus design temperatures so that the vessel walls may be made slightly less thick. Alternatively, the Nuclear Section of the ASME Code may be used to effect possible savings. These and similar points need to be carefully studied along with studies of the operating flexibility of the final design.

Table CS-6.7   Cost Comparison (1972 Costs)

Basis: 3 yr	150 Atm	225 Atm	300 Atm
Converter	$756,000	$626,500	$721,700
Fresh feed and recycle compressors and turbines	743,200	874,500	896,200
Refrigeration compressors and turbines	361,700	310,100	286,300
Catalyst cost[a]	228,700	124,440	103,250
Sub-total	2,089,600	1,935,540	2,007,450
Misc. equip cost difference (piping, etc.)[b]	—	25,000	32,000
Fuel cost difference[c] (Basis: 3 years, 40¢/MM BTU)	—	133,923	93,746
Net difference	—	4,863	43,596

[a] $0.333/lb.

[b] Incremental increase due to higher pressure for piping, heat exchangers, drums, etc.

[c] From Tables CS-6.5 and CS-6.6. Operating factor = 0.98.

Operating Flexibility

A designer can be much more confident in the operability of his selected design by using the "what if" approach, discussed on p. 264[1]. For example, if the catalyst activity factor, ψ, in the first bed should decline to 0.5 and in the other beds to 0.8 how should production be maintained.

Questions such as this are answered using a revised design program which can be called an operating program. Briefly, now the inputs become catalyst quantities (1.429 times amounts shown in Table CS-6.1), inlet temperatures, feed composition, and pressure. We desire an output ammonia production and bed outlet temperatures. To simplify the program, the separator loop is eliminated and the inert level in the recycle fixed at an average value. This does not affect the outcomes greatly and enables the simplified program to be incorporated into an optimization routine if desired.

The answer to the question is given in Table CS-6.8 which indicates that the unit as designed has good flexibility. Other similar concerns can be resolved in this manner with assurance that the conclusions are reasonably correct.

Table CS-6.8   Results of Operating Study (150 Atm)

Bed No.	Temp. In °F	Temp. Out °F	NH_3 In %	NH_3 Out %
1	850	959	5.18	9.54
2	750	853	8.43	12.69
3	750	814	11.58	14.29
Outlet flow, moles/hr				62,882.5
Ammonia production, tons/day				1114[a]

[a] The design production can clearly be exceeded even with deactivated catalyst, but the total flow must be increased which means higher recycle compressor costs.

REFERENCES

1. A. Nielsen, *Catalysis Reviews*, **4** (1), 1–26 (1970).
2. C. L. Thomas, *Catalytic Processes and Proven Catalysts*, Academic, New York, 1970.
3. D. C. Dyson and J. M. Simon, *Ind. Eng. Chem. Fundam.*, **7**, 605 (1968).
4. A. Nielsen, J. Kjaer, and B. Hansen, *J. of Catal.*, **3**, 68 (1964).
5. L. J. Gillespie and J. A. Beattie, *Phys. Rev.*, **36**, 743 (1930).
6. H. W. Cooper, *Hydrocarbon Process.*, **46** (2), 159 (1967).
7. R. H. Newton, *Ind. Eng. Chem.*, **27**, 302 (1935).
8. H. R. Shaw and D. R. Wones, *Am. J. Sci.*, **262**, 918 (1964).
9. G. Guerreri, *A.I.Ch.E.J.*, **13**, 877 (1967).

10. G. Guerreri, *Hydrocarbon Process.*, **49** (12), 74 (1970).

11. M. J. Shah, *Ind. & Eng. Chem.*, **59** (1), 72 (1967).

12. B. H. Sage, R. H. Olds, and W. N. Lacey, *Ind. Eng. Chem.*, **40**, 1453 (1948).

13. O. J. Quartulli, J. B. Fleming, and J. A. Finneran, *Hydrocarbon Process.*, **47** (11), 153 (1968).

14. A. Michels, T. Wassenaar, G. J. Wolkers, W. deGraaff, and P. Louwerse, *Appl. Sci. Res.*, **A3**, 1 (1951).

15. A. Nielsen, *Investigation on Promoted Iron Catalyst for the Synthesis of Ammonia*, 3rd ed., Gjellerups, Copenhagen, 1968.

CASE STUDY 107

Sulfur Dioxide Oxidation

THERE EXISTS no better opportunity than SO_2 oxidation to demonstrate the complexity and problems that confront the designer in developing adequate and meaningful design models. Superficially, the system presents the classically simple case of an adiabatic single reaction with negligible side reaction. Precise physical and thermodynamic data exist (10).

$$SO_2 + \tfrac{1}{2}O_2 \rightleftharpoons SO_3$$

As will be seen, however, the reaction is mechanistically complex, exhaustive studies have produced no general agreement on acceptable rate forms, and the active catalytic component is in the molten state at operating conditions. This latter fact can cause the apparent effective diffusivity to change markedly with temperature depending on the way the liquid distributes within the pores (see p. 148[1]).

Problem Statement

The following are typical data obtained from a commercial sulfuric-acid plant converter using sludge acid as the feed to the burner that precedes the converter.

Reactor diameter: 35 ft

Feed composition to converter:

Component	Mole %
SO_2	6.26
O_2	8.30
CO_2	5.74
N_2	79.70
	100.00

Feed rate: 10,858 lb moles/hr

Bed No.	Temperature, °F		Exit Conversion	Catalyst Height, ft	Inlet Pressure In. H_2O
	In	Out			
1	867	1099	68.7	1.276	63
2	851	923	91.8	1.408	51
3	858	869	96.0	1.511	41
4	815	819	97.5	1.848	36

Develop a model useful for design purposes that will predict bed heights and produce reasonable agreement with this and other similar operating data.

Catalyst Properties

	Wt. %	Mole Ratio Comp/V_2O_5
V_2O_5	8.2	
K_2O	12.3	2.9
Na_2O	1.2	0.43
Fe_2O_3	1.0	

$S_g = 1-2 \text{ m}^2/\text{g}$ $\rho_b = 0.567 \text{ g/cc}$
$V_p = 0.35 \text{ cc/g}$ $\rho_p = 1.1729 \text{ g/cc}$
Size 0.22 in. × 0.40 in. cylinders

Kinetics and Mechanism

The many rate equations that have been proposed over the years for this interesting vanadium pentoxide catalyst have been reviewed (1–3,8,14). Most of the commercial catalysts are supported on some type of siliceous material, such as kieselguhr. Variations in preparing the support can account for much of the apparent inconsistency in rate data gathered on pelleted catalysts. Quite obviously, the effectiveness factors of different forms of apparently similar catalysts can vary greatly. More recent studies using finely ground catalyst have eliminated intraparticle diffusion as a variable and brought more consistency in the rate data and mechanistic arguments. It is now generally agreed that the mechanism involves oxidation–reduction of active V_2O_5 that exists on the support at operating conditions in the molten state (3–5,14).

Various schemes have been proposed. The following is typical (9).

$$SO_2 + 2V^{5+} + O^{2-} \rightleftharpoons SO_3 + 2V^{4+} \qquad (1)$$

$$\tfrac{1}{2}[O_2 + V^{4+} \rightleftharpoons V^{5+} + O_2^{-}] \qquad (2)$$

$$\tfrac{1}{2}[O_2^{-} + V^{4+} \rightleftharpoons V^{5+} + 2O^{-}] \qquad (3)$$

$$O^{-} + V^{4+} \rightleftharpoons V^{5+} + O^{2-} \qquad (4)$$

Various rate equations can be derived based on this mechanistic scheme and the designated rate-controlling step. Using fine catalyst, for which the effectiveness factor is unity, it has been shown that the rate equation based on Eq. 3 being the slow step with 1, 2, and 4 at equilibrium fits the data most successfully in the range above 30% conversion (9).

Although this equation correlates data in the range above 30% conversion very well, it is not satisfying for modeling because it is indeterminant at the limit of zero partial pressure of SO_3. An alternate equation that does not suffer this weakness involves similar arguments, but uses a different mechanism (5).

With reactions 2 and 3 combined ($\tfrac{1}{2}O_2 + V^{4+} \overset{k_5}{\rightleftharpoons} V^{5+} + O^{-}$) as rate controlling and reactions 1 and 4 at equilibrium, a straightforward derivation yields (3,5,9)

$$(-\hat{r}_{SO_2}) = \psi \eta \hat{k}_{p_m} \frac{K_M P_{SO_2}/P_{SO_3}}{[1 + (K_M P_{SO_2}/P_{SO_3})^{\frac{1}{2}}]^2} \left[P_{O_2} - \left(\frac{P_{SO_3}}{P_{SO_2} K_p} \right)^2 \right] \qquad \text{(CS-7.1)}$$

or

$$(-\hat{r}_{SO_2}) = \psi \eta \hat{k}_{p_m} \frac{K_M P_{SO_2}}{[P_{SO_3}^{\frac{1}{2}} + (K_M P_{SO_2})^{\frac{1}{2}}]^2} \left[P_{O_2} - \left(\frac{P_{SO_3}}{P_{SO_2} K_p} \right)^2 \right] \qquad \text{(CS-7.2)}$$

Rate units are g moles SO_2 conv/(g cat.)(hr), $K_M = K_1 C_{O_2-}$, and $\hat{k}_{p_m} = k_5 C_V^2$, where C_V^2 is the total vanadium concentration. Equation CS-7.1 fits the same experimental data with satisfactory precision (7) and has recently been demonstrated in a separate study on fine catalyst (14) to fit data over a wide temperature range better than any of the eleven other equations tested.

The value of K_M is agreed to be somewhat insensitive to catalyst receipe in the low range (3,5,7). But a study of a variety of catalysts with varying Na_2O content indicates Na_2O content to be an important variable related to $\hat{k}_{p_m}$, as shown in Fig. 1.5, p. 27[1], (7). Conveniently, these data for fine catalyst of unit effectiveness factor exhibit an isokinetic point. Thus E is a function of A, and the best fit for the data can be selected by trying different reasonable values of E; From Fig. 1.5, $\ln A = 7.108707 \times 10^{-4} E - 1.365433$. The emphasis on reasonable is essential to prevent any such effort from being a curve fitting expedition glossed over with a veneer of theory. For the model to have predictive value, it must be consistent with physical reality. In this case that means E should lie somewhere between curves 1 and 2 ($E = 41{,}400$–$53{,}800$), which is in the range of Na_2O content of interest. Interpolating for the catalyst in question $E \cong 47{,}000$ cal/(g-mole)($^\circ$K).

Effectiveness Factor

The complexity of the rate form suggests the need for deriving an effectiveness factor based on this form. A rather thorough study of effectiveness factors for the type catalyst being considered has been presented, but a simpler rate form applicable over modest ranges of conversion was used and an effectiveness factor derived as follows (6,12,13).

$$(-\hat{r}_{SO_2}) = \hat{k}_p\left(P_{SO_2} P_{O_2}^{\frac{1}{2}} - \frac{P_{SO_3}}{K_p} \right) \tag{CS-7.3}$$

Mass-Transfer Balances for Spherical Particle ($P = 1$ Atm)

$$\frac{d}{d\mathbf{r}_I}\left(\mathbf{r}_I^2 \mathscr{D}_{ISO_2} \frac{dy_{SO_2}}{d\mathbf{r}_I} \right) - \mathbf{r}_I^2(-\hat{r}_{SO_2})RT\,\rho_p = 0 \tag{CS-7.4}$$

$$\frac{d}{d\mathbf{r}_I}\left(\mathbf{r}_I^2 \mathscr{D}_{IO_2} \frac{dy_{O_2}}{d\mathbf{r}_I} \right) - \mathbf{r}_I^2\tfrac{1}{2}(-\hat{r}_{SO_2})RT\,\rho_p = 0 \tag{CS-7.5}$$

$$\frac{d}{d\mathbf{r}_I}\left(\mathbf{r}_I^2 \mathscr{D}_{ISO_3} \frac{dy_{SO_3}}{d\mathbf{r}_I} \right) + \mathbf{r}_I^2(-\hat{r}_{SO_2})RT\,\rho_p = 0 \tag{CS-7.6}$$

where $\mathbf{r}_I$ is the radial distance in an equivalent sphere.

From stoichiometry and Eqs. CS-7.4–CS-7.6, expressions for y_{SO_3} and y_{O_2} can be obtained in terms of y_{SO_2} upon integrating.

$$y_{O_2} = (y_{O_2})_s + \frac{1}{2}\frac{\mathscr{D}_{ISO_2}}{\mathscr{D}_{IO_2}}\left[y_{SO2} - (y_{SO_2})_s \right] \tag{CS-7.7}$$

$$y_{SO_3} = (y_{SO_3})_s - \frac{\mathscr{D}_{ISO_2}}{\mathscr{D}_{ISO_3}}\left[y_{SO_2} - (y_{SO_2})_s \right] \tag{CS-7.8}$$

where suffix s indicates a value measured at the exterior surface of the catalyst. Then Eq. CS-7.4 can be rewritten using Eqs. CS-7.7 and CS-7.8, conversion X, and assuming an isothermal particle

$$\frac{d^2X}{d\mathbf{r}_k^2} + \frac{2y_0}{2 - y_0 X}\left(\frac{dX}{d\mathbf{r}_k} \right)^2 + \frac{2}{\mathbf{r}_k}\frac{dX}{d\mathbf{r}_k} = -\frac{D_p^2 RT(2 - y_0 X)^2}{4(4y_0 - 2y_0^2)\mathscr{D}_{ISO_2}}\,\rho_p(-\hat{r}_{SO_2}) \tag{CS-7.9}$$

where y_0 is the initial mole fraction SO_2 in feed and $\mathbf{r}_k = 2\mathbf{r}_I/D_p$ similarly values of partial pressures in Eq. CS-7.3 be written in terms of y_0 and X.

Equation CS-7.9 has been solved numerically (13) and $(dX/d\mathbf{r}_k)\mathbf{r}_k = 1$ evaluated so that the effectiveness factor can be obtained.

$$\eta = \frac{\text{observed rate}}{\text{rate based on concentrations @ exterior surface}}$$

$$= \frac{6y_0(2 - y_0)}{(2 - y_0 X_s)^2}\left(-\frac{dX}{d\mathbf{r}_k}\right)_{\mathbf{r}_k = 1} \frac{1}{\phi_m f(X_s)} \tag{CS-7.10}$$

where

$$\phi_m = \frac{D_p{}^2 RT\,\hat{k}_p \rho_p}{4\mathscr{D}_{I\mathrm{SO}_2}3600} = 9\left(\frac{V_k}{a_p}\right)^2 \frac{RT\,\hat{k}_p \rho_p}{(\mathscr{D}_{I\mathrm{SO}_2})(3600)} \tag{CS-7.11}$$

$f(X)_s$ is the rate equation in X and y_0 without $\hat{k}_p$; $\hat{k}_p$ is the rate constant based on Eq. CS-7.3, g moles/(g cat.)(hr)(atm)$^{1.5}$, V_k is the particle volume; and $R = 82.06$. The RT term converts $\hat{k}_p$ to required concentration units.

The solution was rather insensitive to conversion and feed composition, and a convenient empirical formula for η as a function of ϕ_m was obtained from the calculated data.

$$\eta = (\phi_m + C_1)/(A_1\phi_m + B_1) \qquad \text{for } 3 < \phi_m < 400 \quad \text{(CS-7.12)}$$
$$A_1 = 8.52518$$
$$B_1 = 539.706$$
$$C_1 = 503.004$$

$$\eta = A_2(\phi_m)^{B_2} \qquad \text{for } \phi_m > 400 \tag{CS-7.13}$$
$$A_2 = 3.8299$$
$$B_2 = -0.46748$$

Since all this work has been already accomplished, it seems reasonable to apply it directly in the computational program, by calculating the rate at any point first using Eq. CS-7.1 with $\eta = 1.0$. Then equate that rate to Eq. CS-7.3 and solve for $\hat{k}_p$ in Eq. CS-7.3 for use in Eqs. CS-7.11–CS-7.13.

An effective diffusivity for use in Eq. CS-7.9 was determined from the extensive experimental data reported on a catalyst similar to the subject catalyst (12).

$$\mathscr{D}_{I\mathrm{SO}_2} = 0.0286 \text{ cm}^2/\text{sec}$$

Reactor Model

This reactor is a typical multibed adiabatic reactor with intermediate cooling. Given good thermodynamic data, which are existent for this system (10), it should be possible to calculate adiabatic temperatures for the observed

ALGORITHM 91

outlet conversions that agree with measured values for each bed provided temperatures and compositions are determined accurately. Since in the first bed, particularly, catalyst surface temperatures will be greater than bulk temperatures, thermocouples must not touch the catalyst. A location just above the bed on the inlet and at the inert support-catalyst interface at the outlet have been recommended (11). Precise methods for analysis are also suggested (11).

Since SO_2 converters operate near atmospheric pressure and do not require high velocities for exterior heat transfer, it is only necessary to minimize pressure drop in order to minimize power consumption. Thus these converters are designed for low mass velocities with only about $\frac{1}{2}$–1 in. of H_2O ΔP/ft of bed. For this reason, unlike high-pressure adiabatic reactors where larger ΔPs are permissible, interfacial gradients between the catalyst surface and the bulk phase will occur in the first bed because of high rates. Accordingly, equations for this phenomenon will be included.

Design Equations

Basis: 1 mole of feed

$$\frac{dn_{SO_2}}{dZ} = \frac{\rho_b(-\hat{r}_{SO_2})M_F}{G} \quad \text{or} \quad \Delta Z = \frac{\Delta n_{SO_2}G}{\rho_b(-\hat{r}_{SO_2})M_F}$$

$$d(\Sigma n_j H_j) = (-\Delta H_{SO_2})dn_{SO_2} \quad \text{or} \quad \Delta n_{SO_2} = \frac{\Sigma n_j \Delta H_j}{(-\Delta H_{SO_2})}$$

where n_j is the moles of component j per mole of total feed and H_j is the enthalpy of any component j.

$$k_{g_j}{}^s a_m[(P_j)_b - (P_j)_s] = (-\hat{r}_{SO_2})$$

$$ha_m(T_s - T_b) = (-\hat{r}_{SO_2})(-\Delta H_{SO_2})$$

where $a_m = a_p/V_k\rho_b$, b indicates bulk conditions, and s indicates surface conditions.

Algorithm

1. Select a temperature increment (1°F was found satisfactory).
2. Calculate ΔH_{SO_2} @ $T_n + \Delta T$

$$\Delta H_{SO_2} = a + bT + cT^2 + dT^3 *$$

$a = 4.1923286 \times 10^4$ $b = -64.3951192$

$c = 7.52214287 \times 10^{-2}$ $d = -2.94166667 \times 10^{-5}$

* Based on curve fits of tabular data (10).

3. Calculate H_j's @ T_n and $T_n + \Delta T$

$$H_j = a' + b'T + c'T^2 + d'T^3$$

Basis: enthalpy above $298°K$*

	a'	b'
SO_2	-2.651×10^3	7.41333334
SO_3	-3.4906571×10^3	9.16952383
O_2	-1.8469429×10^3	5.6429762
N_2	-1.9181143×10^3	6.27571429
CO_2	-2.0802286×10^3	5.59059524

c'	d'
4.8×10^{-3}	$-1.33333333 \times 10^{-6}$
$7.73571428 \times 10^{-3}$	$-2.1666666 \times 10^{-6}$
$2.21428571 \times 10^{-3}$	-5.833333×10^{-7}
$7.71428566 \times 10^{-4}$	$2.3925238 \times 10^{-15}$
$5.89285714 \times 10^{-3}$	$-1.4166667 \times 10^{-6}$

4. Calculate $\Sigma n_j \Delta H_j = \Sigma n_j (H_{j_{T_{n+\Delta T}}} - \Delta H_{j_{T_n}})$

5. Calculate $\Delta n_{SO_2} = \dfrac{\Sigma n_j \Delta H_j}{(-\Delta H_{SO_2})}$

6. Calculate conversion $X_{n+1} = \dfrac{(n_{SO_3})_n + \Delta n_{SO_2}}{(n_{SO_2})_0}$

7. Calculate moles each component per mole of feed

$$(n_{SO_2})_{n+1} = (n_{SO_2})_n - \Delta n_{SO_2}$$

$$(n_{SO_3})_{n+1} = (n_{SO_3})_n + \Delta n_{SO_2}$$

$$(n_{O_2})_{n+1} = (n_{O_2})_n - \tfrac{1}{2}\Delta n_{SO_2}$$

$$(n_T)_{n+1} = (n_{SO_2})_{n+1} + (n_{SO_3})_{n+1} + (n_{O_2})_{n+1} + (n_{CO_2})_0 + (n_{N_2})_0$$

8. Calculate mole fractions

$$(y_{SO_2})_{n+1} = \frac{(n_{SO_2})_{n+1}}{(n_T)_{n+1}}, \text{ etc.}$$

* Based on curve fits of tabular data (10).

ALGORITHM 93

9. Calculate average mole fractions for increment and average molecular weight (M_m)

$$\bar{y}_{SO_2} = \frac{(y_{SO_2})_n + (y_{SO_2})_{n+1}}{2}, \text{ etc.}$$

$$M_m = \Sigma \bar{y}_j MW$$

10. Calculate average bulk partial pressures for increment

$$\bar{P}_{SO_2} = \bar{y}_{SO_2} P, \text{ etc.}$$

11. (a) Calculate rate based on known conditions using Eq. CS-7.1 with $\psi = 1$ and $\eta = 1$.

$$\hat{k}_{p_m} = \exp\left(A - \frac{E}{R'T}\right), \frac{\text{g moles SO}_2}{\text{(g cat.)(hr)(atm)}}$$

$$\ln A = 7.108707 \times 10^{-4} E - 1.365433$$

$$K_m = 2.3 \times 10^{-8} \exp\left(\frac{27,200}{R'T}\right)$$

$$\log_{10} K_p = \frac{5.14488992 \times 10^3}{T} - 4.8882412*$$

η is determined from Eqs. CS-7.11–CS-7.13.
(b) Calculate value of $\hat{k}_p$ from

$$(-\hat{r}_{SO_2}) = \hat{k}_p\left(P_{SO_2} P_{O_2}^{\frac{1}{2}} - \frac{P_{SO_3}}{K_p}\right)$$

(c) Calculate ϕ_m and η
(d) Calculate new value of rate by correcting original value by multiplying by η.

12. Calculate surface temperature

$$T_s = \frac{(-\hat{r}_{SO_2})(-\Delta H_{SO_2})}{ha_m} + T_b$$

If $T_s - T_b = 2$ or less, go to step 15. If greater than 2, proceed to step 13.
13. Calculate surface partial pressures

$$(P_j)_b - (P_j)_s = \frac{(n)(-\hat{r}_{SO_2})}{k_{g_j}{}^s a_m}$$

$n = 1$ for SO_2, 0.5 for O_2, and -1 for SO_3.

___
* Curve fit from Ref. 10.

14. Go back to step 11.
15. Calculate ΔZ and Z

$$\Delta Z = \frac{\Delta n_{SO_2} G}{\rho_b(-\hat{r}_{SO_2})M_F}$$

16. Go to desired conversion. Then proceed to next bed.

Operating inlet pressure was used at the inlet of each bed and pressure was corrected for ΔP loss for each increment using Eq. 11.8B. With appropriate values of the parameters for catalyst. Values for $k_{g_j}{}^s$ and h were obtained using Eqs. 11.12 and 11.15–16 with $N_{Pr}{}^{\frac{1}{3}} = 0.8$ and $(N_{Sc})^{\frac{1}{3}} = 1.334$, 1.272, and 1.041 for SO_3, SO_2, and O_2, respectively. Viscosities of the mixture were determined from

$$\mu_m = \frac{\sum\limits_{j=1}^{n} [\bar{y}_j \mu_j (M_j)^{0.5}]}{\sum\limits_{j=1}^{n} [\bar{y}_j (M_j)^{0.5}]}$$

$$\mu_j = \exp(\bar{A} + \bar{B}T)$$

	$\bar{A}$	$\bar{B}$
SO_2	-5.012	0.0020196
SO_3	-4.855	0.0020196
O_2	-4.172	0.001213
N_2	-4.65	0.002032
CO_2	-4.0571	8.57×10^{-4}

Results

Because of the lack of experimental data for the catalyst used in the commercial reactor, an effective diffusivity was determined using experimental reaction-rate data reported in the literature (12). This value (0.286 cm^2/sec), though based on data obtained in the temperature range of 860–968°F, was used for all four beds. The predicted lengths for beds 1 and 4 were not acceptable (Table CS-7.1) although values for bed 2 and 3 are reasonable. The bed length or catalyst volume is the more sensitive parameter, and all calculations were made to the observed conversion for each bed. If one calculates to the known length, the values of conversion thus obtained may appear rather close to the observed value and give a false sense of security concerning the efficacy of the model.

Experimental values of apparent effective diffusivities, which were determined for a different V_2O_5 catalyst than used for rate data, are in the range of bed 4 operating temperature and are approximately 35% of those at

Table CS-7.1 Comparison of Model Predictions with Operating Data[a]

No.	Actual Bed Depth, Ft	Operating Temp., °F In	Operating Temp., °F Out	Case 1 Conv. %	Case 1 $\mathcal{D}_{ISO_2}$	Case 1 Temp. Out, °F	Case 1 Bed Depth Ft	Case 2 $\mathcal{D}_{ISO_2}$	Case 2 Temp. Out, °F	Case 2 Bed Depth, Ft	Case 3 $\mathcal{D}_{ISO_2}$	Case 3 Temp. Out, °F	Case 3 Bed Depth, Ft
1	1.276	867	1099	68.7	0.0286	1090	0.629	0.027	1090	0.641	0.025	1090	0.657
2	1.408	851	923	91.8	0.0286	926	1.518	0.027	927	1.559	0.025	927	1.614
3	1.511	858	869	96.0	0.0286	872	1.431	0.027	872	1.461	0.025	872	1.523
4	1.848	815	819	97.5	0.0286	820	1.191	0.011	820	1.848	0.011	820	1.848

[a] $E = 47,000$.
Temperature into each bed and conversion in and out set at values observed in operating plant.

higher temperatures such as in beds 1–3. Accordingly, it was decided to determine the appropriate value for bed 4 by selecting the one that caused the operating data to be reproduced. Because bed 4 also operates very close to the isokinetic point, it is rather insensitive to the energy of activation used. Thus the best-fit-value of $\mathscr{D}_{I_{SO_2}} = 0.011$ cm^2/sec given in case 3 of Table CS-7.1 should be a rather accurate representation of the apparent diffusivity at operating conditions, and it is 38 % of that of 0.0286 determined in the 860–968°F range in a separate study (12).

Using the same technique on bed 3 a value of $\mathscr{D}_{I_{SO_2}} = 0.025$ cm^2/sec was determined. This value is very close to the value of 0.0286 determined from laboratory reaction rate data from what appears to be a similar catalyst (12). Clearly effective diffusivities in the range 0.025–0.029 are acceptable for beds 2 and 3. There exists no independent evidence that the effective diffusivities at the conditions of beds 1, 2, and 3 would differ, and the value of $\mathscr{D}_{I_{SO_2}}$ determined in bed 3 was thus also used for beds 1 and 2. The results of the calculations are summarized in Tables CS-7.1–CS-7.3.

The effect of changing both $\mathscr{D}_{I_{SO_2}}$ and inlet temperatures are illustrated. It is important to recognize that the reported operating temperatures could be in error by 5–10°F. Referring to Table CS-7.3 the effects of small changes in inlet temperature can be rather dramatic depending on the proximity to equilibrium. The temperature sensitivity of bed 2 is seen to be much less

Table CS-7.2   Effect of Inlet Temperature on Bed Depth

Bed No.	Inlet Temperature °F	Calculated Height, Ft
1	857	0.684
	867	0.657
	877	0.644
2	841	1.634
	851	1.614
	855	1.600
	861	1.607
3	848	1.393
	858	1.523
	868	2.040
4	805	1.724
	815	1.848
	825	2.102

$\mathscr{D}_{I_{SO_2}}$ same as for Case 3, Table CS-7.1.

Table CS-7.3   Effect of Possible Errors in Plant Observations

| | | | | | Altered Conditions | | | |
| | | | | | | | | |
Bed No.	Actual Bed Depth, Ft	Operating Temp., °F In	Out	Conv. %	Temp. In °F	Calc. Temp. Out, °F	Conv. %	Bed Depth, Ft
1	1.276	867	1099	68.7	867	1099	71.65	0.748
2	1.408	851	923	91.8	857	923	91.85	1.487
3	1.511	858	869	96.0	855	869	96.0	1.459
4	1.848	815	819	97.5	815	820	97.5	1.848

$\mathscr{D}_{I_{SO_4}}$ same as for Case 3, Table CS-7.1.

than bed 3. The model, once established, could be useful in determining the optimum inlet temperature for each bed.

Table CS-7.3 is the result of some speculative calculations in which the outlet conversion of the first bed was changed to yield the measured outlet temperature, and the inlet temperatures of beds 2 and 3 were altered so that the measured outlet temperatures are obtained. Good agreement again results for beds 2, 3, and 4; but bed 1, though closer to the actual catalyst loading, is still 41% off. It is reasonable to hypothesize that bed 1 becomes partially deactivated during the early hours of start-up. This is not uncommon in the first bed of exothermic adiabatic reactors. This hypothesis must be tested by independent studies of the catalyst.

Improved rate equations, which are appearing in the literature, should be tested. In fact, it has more recently been shown that a simplified model involving liquid phase diffusion in the vanadium oxide melt and homogeneous reaction is satisfactory for the range of temperatures covered by the first three beds in this example (15). In the low temperature range corresponding to the range of the fourth bed ($< 435°C$), preliminary experimental evidence suggests rather profound changes in catalyst composition when passing between kinetic and diffusion-controlled regimes (15). We are realizing that SO_2 oxidation, which initially appeared to be a rather simple reaction system, is really quite complex; and it has become a dramatic illustration of the many difficulties in modeling reactions.

REFERENCES

1.   S. Weychant and A. Urbanek, *Int. Chem. Eng.*, **9**, 396 (1969).
2.   G. Honti, *Annales du Genie Chimique, Congress Int'l. du Soufre*, Toluse, May 22–26, 1967.
3.   P. Mars and J. G. H. Maessen, *J. Catal.*, **10**, 1 (1968).

4. A. Simecek, B. Kadlec, and J. Michalek, *J. Catal.*, **14**, 287 (1969).
5. P. Mars and J. G. H. Maessen, 3rd International Congress on Catalysis, Vol. 1, p. 266, Amsterdam, 1964.
6. B. Kadlec, J. Michalek, and A. Simecek, *Chem. Eng. Sci.*, **25**, 319 (1970).
7. A. Simecek, *J. Catal.*, **18**, 83 (1970).
8. S. Minhas and J. J. Carberry, *Brit. Chem. Eng.*, **14** (6), 799 (1969).
9. Regner, A. and A. Simecek, *Coll. Czech. Chem. Comm.*, **33**, 2540 (1968).
10. JANAFF *Thermochemical Tables*, 2nd Ed. National Bureau of Standards, Washington, D.C., 1971.
11. H. Z. Hurlburt, *Monitoring Contact Acid Plant Converters*, 71st American Institute of Chemical Engineers, National Meeting, Dallas, Tex., Feb. 20–23, 1972.
12. B. Kadlec and V. Pour, *Coll. Czech. Chem. Comm.*, **33**, 2526 (1968).
13. B. Kadlec and A. Regner, *Coll. Czech. Chem. Comm.*, **33**, 2388 (1968).
14. H. Livbjerg and J. Villadsen, *Chem. Eng. Sci.*, **27**, 21 (1972).
15. H. Livbjerg, K. F. Jensen, and J. Villadsen, *J. Catal.*, to be published (1976).

CASE STUDY 108

Catalytic Reforming

THIS CASE STUDY illustrates the effective use of pseudocomponents to simplify a complex reaction system with a feed containing numerous chemical entities. Again adiabatic-staged reactors are specified the first two of which involve net endothermic reactions and the last, a net exothermic effect. Reheating between stages is necessary. More detailed models than that used here have been described that include other important reactions such as deactivation by coking (e.g., Ref 4). The data for excellent studies of this type remain proprietary.

Problem Statement

Design a catalytic reforming unit for producing 20,000 BPSD of C_5+ reformate having a research octane number of 95.

Typical Hydrodesulfurized Feedstock Analysis

Gravity: 51.9 API

ASTM Distillation:

IBP	240°F	60%	302°F
10%	262	70	311
20	272	80	322
30	279	90	335
40	284	95	346
50	292	EP	369

Compound Type:

	Mole %
paraffins	39.4
naphthenes	40.7
aromatics	19.9
	100.0

molecular weight: 123; characterization factor: 11.7

VABP: 293°F

MABP: 285°F

Received at battery limits at 40 psig and 200°F.

Catalyst

The catalyst is a 0.6% Pt-on-alumina ($\frac{1}{16}$ in. × $\frac{3}{16}$ in. extrudate) with $\rho_b = 0.78$ g/cm^3, $\rho_p = 1.2$ g/cm^3, $a_m = 471$ m^2/g, and pore volume = 0.42 cm^3/g. Costs in 1968 were \$13.80/lb with an approximate royalty of \$35/bbl of capacity, and a recovery cost of \$3.25/lb for new support. In addition to the poisons shown in Table 2.13, coking deactivates the catalyst but can be removed by regenerating with air. The catalyst has a dual-function with an acidic function for carbonium-ion reactions and a metallic function for hydro-dehydrogenation reactions. The acidic characteristics are thought to occur by partial replacement of OH on the alumina and the corresponding enhancement of activity of the residual hydrogen atoms upon treatment with halogens or halogen compounds (1).

Chemistry

Dual-function catalysts, both Pt-on-alumina and Pt-on-silica-alumina, have been studied extensively using various pure compounds as typical examples of constituents in refinery naphtha streams (1). An overall reaction scheme is given in Fig. CS-8.1 for C_6 hydrocarbons, showing the acidic functions horizontally and the hydro-dehydrogenation function vertically. In addition to these reactions, hydrocracking (e.g., $C_7H_{16} + H_2 \rightarrow C_4H_{10} + C_3H_8$) and coking are also important. Isomerization of straight-chain to branched paraffins, aromitization, and hydrocracking each enhance the octane number of the product, but aromitization plays the major role because of the high octane numbers for aromatics (in the range of 120–140). Some hydrocracking is often essential, however, to meet high octane requirements.

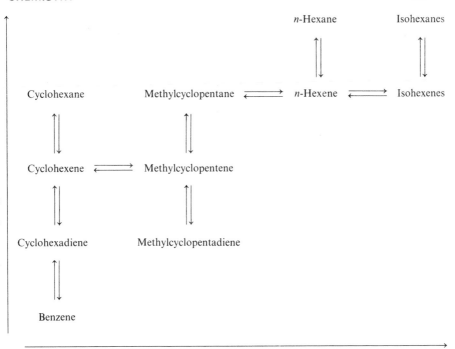

Fig. CS-8.1 Reaction paths for reforming C_6 hydrocarbons on dual-function catalysts. Reproduced by permission: J. H. Sinfelt, *Adv. Chem. Eng.*, **5**, 38 (1964). [Copyright by Academic Press.]

It breaks up large low-octane fragments into higher-octane molecules of smaller size. More importantly, it increases octane number by increasing the concentration of aromatics simply by reducing the specific volume of the nonaromatic fraction remaining in the liquid after flashing and distilling the reactor product. Octane number has been correlated with volume percent aromatics (VPAR) on the C_5 + reformate and is 58% for the 95-octane goal of this study (2). The effect of paraffin octane numbers (in the range of 40–60) is not significant.

Coking deactivates the catalyst, but it can be minimized by rapid hydrogenation of unsaturated coke precursors. Thus coking is prevented to a large extent by using high hydrogen partial pressures—high H_2/oil ratios and/or high pressures. Continuous catalyst use before regeneration of up to 26 mo has been reported (3). Newer forms of this catalyst will exceed this performance even with lower hydrogen partial pressures. Coking characteristics are always difficult to quantify, and the designer must often depend on pilot-plant and plant operating data. Assume such data indicate that in the range of 425–575 psig that 1 yr of operating without regenerating is possible with a pure hydrogen-to-oil ratio of 5.9, which is a typical reported value (2). The catalyst activity at the end of the run will be 20 % of the virgin activity; and to compensate for declining activity, operating temperature is raised gradually to a maximum allowable based on end-of-run conditions. Although aromatics yield will be improved at inlet pressures below 475 psig, coking will increase; but the data presented here do not define the lower pressure region. As coking rate increases it ultimately will become necessary to regenerate more frequently and require an extra or swing reactor. This mode of operation will be avoided by restricting the minimum pressure to 475 psia.

The complex feed mixture and product stream can be greatly simplified by considering four model reactions that are crucial in determining yield* and octane number (2).

$$\text{Naphthenes} \rightleftharpoons \text{aromatics} + 3H_2 \tag{1}$$

$$\text{Paraffins} \rightleftharpoons \text{naphthenes} + H_2 \tag{2}$$

$$\text{Hydrocracking of paraffins} \tag{3}$$

$$\text{Hydrocracking of naphthenes} \tag{4}$$

The C_1 through C_5 fractions produced by hydrocracking were observed experimentally to occur in approximately equal molar portions (2). Thus for paraffin cracking the stoichiometric equation would be, for $C_{15}H_{32}$ as an example

$$C_{15}H_{32} + 4H_2 \longrightarrow C_1 + C_2 + C_3 + C_4 + C_5$$

or for the general case

$$C_nH_{2n+2} + \left(\frac{5n}{15} - 1\right)^\dagger H_2 \longrightarrow$$

$$\frac{n}{15}C_1 + \frac{n}{15}C_2 + \frac{n}{15}C_3 + \frac{n}{15}C_4 + \frac{n}{15}C_5$$

or

$$C_nH_{2n+2} + \left(\frac{n-3}{3}\right)H_2 \longrightarrow$$

$$\frac{n}{15}C_1 + \frac{n}{15}C_2 + \frac{n}{15}C_3 + \frac{n}{15}C_4 + \frac{n}{15}C_5 \quad \text{(CS-8.1)}$$

---

* Yield fraction for a product such as gasoline that is sold by volume is the volume of debutanized reformate produced per volume of liquid feed.

† Number of fragments minus one.

In like manner for naphthene cracking

$$C_n H_{2n} + \frac{n}{3} H_2 \longrightarrow \frac{n}{15} C_1 + \frac{n}{15} C_2 + \frac{n}{15} C_3 + \frac{n}{15} C_4 + \frac{n}{15} C_5$$

$$\text{(CS-8.2)}$$

Each of three hydrocarbon classes—paraffins, naphthenes, and aromatics—can be represented as a single compound having the average properties of that class (2). This reaction system is peculiarly amenable to this type of analysis because the major products have the same number of carbon atoms as the original feed constituents (e.g., heptane $\rightarrow$ methylcyclohexane $\rightleftarrows$ toluene). The nature of close boiling naphtha feed also suggests that each of the hydrocarbon classes in the original feed have the same number of carbon atoms. Hence (2)

$$M_F = n_P M_P + n_N M_N + n_A M_A$$

where n_P, n_N, and n_A designate moles of paraffin ($C_n H_{2n+2}$), naphthene ($C_n H_{2n}$), and aromatic ($C_n H_{2n-6}$) per mole of feed, respectively, n is the number of carbon atoms, and M_F is the molecular weight of the feed.

$$M_F = n_P(12n + 2n + 2) + n_N(12n + 2n) + n_A(12n + 2n - 6)$$

or

$$14n(n_P + n_N + n_A) = M_F - 2n_P + 6n_A$$

or

$$n = \tfrac{1}{14}(M_F - 2n_P + 6n_A) \qquad \text{(CS-8.3)}$$

since $n_P + n_N + n_A = 1$ mole of feed. The average value of n characteristic of a given naptha feed can be determined from Eq. CS-8.3 and was calculated to be 8.82 for the feed in question.

Thermodynamics

The most important reactions are the hydrogenations of cycloparaffins to aromatics. These are rapid reactions that approach equilibrium in a short time. One can often assess complex systems such as this in a qualitative manner by considering a typical pure component such as cyclohexane ($C_6 H_{12} \rightleftarrows C_6 H_6 + 3 H_2$). Calculations for cyclohexane at 500 psia pressure range and H_2-to-oil feed ratio of 5.9 indicate high equilibrium conversions in the range of 900°F.

Extensive experimental reforming runs using naphthas of the type being considered have yielded the following equilibrium constants for naphthenes being converted to aromatics and to paraffins (2):

$$K_{p_1} = \frac{P_A P_H^3}{P_N} = \exp\left(46.15 - \frac{46045}{T}\right), \quad \text{atm}^3 \quad \text{(CS-8.4)}$$

where $T = {}^\circ R$, P_A, P_N, and P_H are the partial pressures in atm of aromatics, naphthenes, and hydrogen, respectively.

$$K_{p_2} = \frac{P_P}{P_N P_{H_2}} = \exp\left(\frac{8000}{T} - 7.12\right), \quad \text{atm}^{-1} \quad \text{(CS-8.5)}$$

where P_P is the partial pressure of paraffins.

When based on moles of H_2 consumed or liberated in a particular reforming reaction, the heat of reaction is essentially independent of molecular weight in the range normally encountered in reforming (2). The values are listed with the corresponding rate equations. The adiabatic factors and heat generation potentials shown in Table 6.4 for the dehydrogenation steps are moderately negative and confirm the need for staged reactors with reheating.

Kinetics

The model reactions 1–4 were used as a basis for empirical rate equations, the constants for which were evaluated from laboratory and plant data (2).

Naphthenes to Aromatic

$$(-\hat{r}_1) = \hat{k}_{p_1}\left(P_N - \frac{P_A P_{H_2}^3}{K_{p_1}}\right), \quad \frac{\text{moles naphthene converted to aromatics}}{\text{(hr)(lb cat.)}}$$

(CS-8.6)

$$\hat{k}_{p_1} = \exp\left(23.21 - \frac{34{,}750}{T}\right), \quad \frac{\text{moles}}{\text{(hr)(lb cat.)(atm)}} \quad \text{(CS-8.6A)}$$

$$\Delta H_1 = 30{,}500 \text{ BTU/mole of } H_2 \text{ liberated}$$

Naphthenes to Paraffins

$$(-\hat{r}_2) = \hat{k}_{p_2}\left(P_N P_{H_2} - \frac{P_P}{K_{p_2}}\right), \quad \frac{\text{moles naphthenes converted to paraffins}}{\text{(hr)(lb cat.)}}$$

(CS-8.7)

$$\hat{k}_{p_2} = \exp\left(35.98 - \frac{59{,}600}{T}\right), \quad \frac{\text{moles}}{\text{(hr)(lb cat.)(atm)}^2} \quad \text{(CS-8.7A)}$$

$$\Delta H_2 = -19{,}000 \text{ BTU/mole of } H_2 \text{ consumed}$$

(Under properly chosen conditions the reverse reaction, paraffins to naphthenes, will become more important.)

Hydrocracking of Paraffins

$$(-\hat{r}_3) = \hat{k}_{p_3} \frac{P_P}{P}, \quad \frac{\text{moles paraffins converted by hydrocracking}}{(\text{hr})(\text{lb cat.})} \quad \text{(CS-8.8)}$$

where P is the total system pressure.

$$\hat{k}_{p_3} = \exp\left(42.97 - \frac{62,300}{T}\right), \quad \frac{\text{moles}}{(\text{hr})(\text{lb cat.})} \quad \text{(CS-8.8A)}$$

$$\Delta H_3 = -24,300 \text{ BTU/mole of } H_2 \text{ consumed}$$

Hydrocracking of Naphthenes

$$(-\hat{r}_4) = \hat{k}_{p_4} \frac{P_N}{P}, \quad \frac{\text{moles naphthenes converted by hydrocracking}}{(\text{hr})(\text{lb cat.})}$$

$$\text{(CS-8.9)}$$

$$\hat{k}_{p_4} = \exp\left(42.97 - \frac{62,300}{T}\right), \quad \frac{\text{moles}}{(\text{hr})(\text{lb cat.})} \quad \text{(CS-8.9A)}$$

$$\Delta H_4 = -22,300 \text{ BTU/mole of } H_2 \text{ consumed}$$

Operating Conditions and Reactor Type

Pressure

We must confine our range of interest to the values (475–575 psig) already presented as acceptable for activity maintenance. The optimum pressure within this range will be determined by a study of catalyst cost, compressor cost, and vessel cost. The reaction is a net producer of hydrogen, and plant hydrogen pressure is not a major factor.

Temperature

Based on pure component data, it is apparent that the operating temperature must be 900°F or above in order to assure negligible reverse reaction and maximize aromatics production. More precise temperatures will be selected on studying the rate characteristics. There will be merit in limiting the maximum reactor temperature to 1000°F in order to safely select a low alloy, 1.25 Cr − 0.5 Mo, steel. This means that the end-of-run temperature should not exceed this temperature. Maximum virgin catalyst temperature is most logically based on the hydrocracking reaction since it is the major cause of catalyst deactivation. Using an activity at the end-of-run of 20% of the virgin activity, the initial maximum temperature will be from Eq. CS-8.8A.

$$(\hat{k}_{p_3})_0 = 0.2\hat{k}_{p_3}$$

or

$$\exp\left(42.97 - \frac{62,300}{T_0}\right) = 0.2 \exp\left(42.97 - \frac{62,300}{1460}\right)$$

Solving for T_0:

$$T_0 = 947 \text{ or } \approx 950°F$$

Thus it appears that we only need explore the range 900–950°F.

Conversion and Yield

Fractional yield for a product sold by volume is most usefully expressed simply as the volume of debutanized reformate per volume of liquid feed. This type of yield does nothing to describe the quality of the product. Quality is specified by a separate statement on the octane number or the % aromatics in the product (58% in this case).

Knowledge of the yield is not essential initially since we can begin with calculations based on W/F only for an adiabatic reactor. Yield data, however, are usually known prior to the design stage from pilot plant studies. For this system a yield of approximately 88% is a reasonable goal, and it is convenient for computer calculations to use preliminary values of component feed rates based on this value which can be adjusted iteratively to give the desired production rate of reformate.

Recycle Ratio

Reactor product gas will be cooled and flashed to separate the hydrogen-rich gases for recycle. For initial calculations one can assume pure hydrogen as recycle, and the composition of the recycle can then be established by successive flash and reactor calculations. For illustrative purposes we will assume a typical recycle composition that has been reported (2) (Mole %: $H_2 = 85.3$, $CH_4 = 8.0$, $C_2H_6 = 5.0$, and $C_3H_8 = 1.7$).

Total recycle ratio: $(5.9)/(0.853) = 6.9$. The preliminary feed material balance is based on 88% yield (see Table CS-8.1).

Reactor Type

Based on the rough adiabatic factor ($-286°F$) for the aromatization reaction it appears that a temperature change of this magnitude would require a

Table CS-8.1 Computer Results for Alternate 1

W/F	Temp. F	Paraffins	Naphthenes	Aromatics	Hydrogen	Methane	Ethane	Propane	Butane	Pentane	VPAR	Yield
					Component Flow Rates, Lb Moles/Hr							
Reactor 1 (P = 475 psia, MW = 20.16)												
0	950	816.56	843.50	412.43	12198.07	1144.02	715.01	243.10	0	0	—	—
6.03	834.2	843.72	376.77	830.16	13359.91	1156.86	727.85	255.94	12.84	12.84	36.67	96.38
Reactor 2												
(Outlet from reactor 1, press. 450 psia)												
0	950	822.10	105.07	1077.38	13987.75	1183.94	754.94	283.03	39.93	39.93	49.15	93.39
Reactor 3												
(Outlet from reactor 2, press. 425 psia)												
0	950	607.50	35.58	1205.81	14130.26	1275.42	846.53	374.51	131.41	131.41	58.08	88.51

Basis: 20,000 BPSD C$_5$+ reformate.

series of adiabatic fixed-bed reactors with intermediate reheating. Reaction 1 is not excessively temperature sensitive, for a 100°F change in temperature produces only a 7.5-fold change in the velocity constant, and temperature drops of 50–100° might be permissable before reheating is necessary.

Design Model

The following equations apply.

Material Balance

$$(-\hat{r}_1)\left(\frac{\Delta W}{F_T}\right) = \Delta X_1 \qquad \text{(CS-8.10)}$$

and similar equations for reactions 2, 3, and 4. X_1 is the conversion in reaction 1, moles naphthenes converted per mole of total feed.

Heat Balance

$$\Delta W\left[(-\hat{r}_1)(-\Delta H_1)(3) + (-\hat{r}_2)(-\Delta H_2) + (-\hat{r}_3)(-\Delta H_3)\left(\frac{n-3}{3}\right) \right.$$

$$\left. + (-\hat{r}_4)(-\Delta H_4)\left(\frac{n}{3}\right) \right] = \sum \mathscr{F}_j c_{p_j} \Delta T \qquad \text{(CS-8.11)}$$

Heat capacities from the *API Data Book* based on nearest true compounds to value of *n* were used.

Products

C$_5$ and lighter hydrocarbons are obtained from the calculated conversions X_3 and X_4 and the stoichiometry expressed by Eqs. CS-8.1 and CS-8.2.

Volume % Aromatics

$$\text{VPAR} = \frac{M_A \mathscr{F}_A/\rho_A}{(M_A \mathscr{F}_A/\rho_A) + (M_P \mathscr{F}_P/\rho_P) + (M_N \mathscr{F}_N/\rho_N) + (M_{C_5} \mathscr{F}_{C_5}/\rho_{C_5})}$$

Volume % Yield

This is simply 100 times the volumetric flow of C$_5$ + reformate divided by the volumetric liquid feed rate all measured at 60°F.

It was assumed that differences between catalyst surface and bulk fluid conditions were negligible. The reactor configuration will be selected to validate this assumption.

The algorithm was similar to that used for Case 102A.

Design Procedure

Preliminary Studies

The exact number of reactors and the operating scheme must be determined by calculating various cases. The several simultaneous and consecutive reactions make qualitative reasoning rather fruitless. In this preliminary stage of process design, however, it is most valuable to obtain an intuition about the variables and their effects on reactor design just as rapidly as is possible for less complex systems. For only by this means of rapidly testing tentative ideas does one obtain useful new approaches. A visual interactive display associated with a digital computer makes such interplay possible. One can, for example, follow with position in the reactor (W/F_T), volume % aromatics and rates of reactions 1 and 3 (aromatics production and hydrocracking of paraffins), which are crucial in setting VPAR and thus octane number.

Hence these two rates together with a plot of the cumulative volume % aromatics, all as a function of W/F were chosen for display. At the end of each trial design the yield was also displayed. A series of preliminary studies by this method using an inlet pressure of 475 psig are shown in sequence in Figs. CS-8.2–CS-8.4, which are line drawings made directly from photographs of the display oscilloscope. If oscilloscope capacity permits, plots of temperature versus W/F_T for each reactor are also useful.

Typical Trial Runs on the Display

Trial 1. Use 900°F inlet and 475 psia pressure and observe behavior in a single reactor. This temperature is too low, the rate of aromatics formation ($\hat{r}_1$) declines too rapidly.

Trial 2. Use 950°F inlet and 475 psia (Fig. CS-8.2). This is a good starting temperature, but it appears that the aromatics production rate declines to a low value such that the first reactor ought to be terminated at $W/F_T \approx 7$. Note also that hydrocracking is negligible.

Trial 3. Use two reactors of same size $W/F_T = 7$ (Fig. CS-8.3; curves 1 and 2). Reheat between reactor 1 and 2 and allow 25 psi for total reheat furnace and reactor pressure drop. Inlet for reactor 1 is 475 psia and 950°F and for reactor 2, 450 psia and 950°F. The volume % aromatics of 49.4 is inadequate, and it is clear that reheating and another reactor are needed.

Trial 4. Add a third reactor of undesignated W/F_T with reheat to 950°F and inlet pressure of 425 psi. (Fig. CS-8.3; curves 1, 2, and 3).

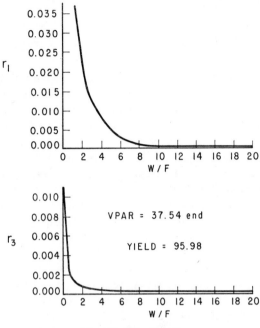

Fig. CS-8.2   Trial 2.

The target percentage aromatics is reached in reactor 3 at $W/F_T \approx 10$. By examining the progress of the rate curves it is apparent that hydrocracking $(\hat{r}_3)$, which occurs primarily in reactor 3, is essential for reaching the desired aromatics content, since the aromatics rate is too low in the third reactor to accomplish this goal by additional conversion to aromatics. It must be done instead by reducing the nonaromatics content of the product through hydrocracking. Trial 4 can be further improved in catalyst utilization by reducing the size of reactors 1 and 2 by a W/F_T of 1, for the last increment of W/F_T is not of much value at the low rate indicated for $\hat{r}_1$ (the volume % aromatics changes imperceptibly in this region in reactor 2).

Trial 5.   (Fig. CS-8.4)

	1	2	3
Inlet temp. °F	950	950	950
Inlet pressure, psia	475	450	425
W/F_T	6	6	10

This seems to be a good basis for beginning detailed design studies. The nature of the display is such that these initial studies are approximate only,

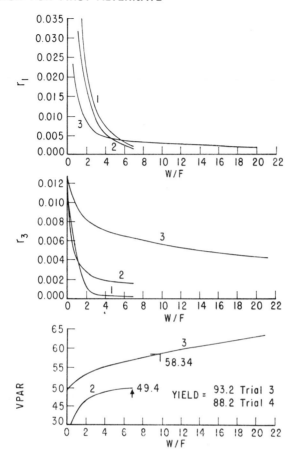

Fig. CS-8.3 Trials 3 and 4.

but at this point the designer has a good feel for the behavior of the variables and has essentially completed the most creative part of the effort, that of selecting the reactor style and arrangement. Further work will involve detailed calculations of various cases including costs of associated equipment in order to obtain the most economical design.

Final Design for First Alternate

The result of computer calculations for the first detailed alternate decided by the preliminary work on the iterative display (Trial 5) are shown in Table CS-8.1. The marked decline in aromitization as the naphthene concentration

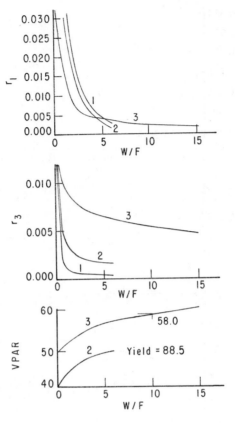

Fig. CS-8.4   Trial 5.

diminishes is evident in the third reactor, as is the corresponding increase in hydrocracking evidenced by the decline in paraffins.

Pressure drop through the reactor beds was considered negligible, but this must be checked on deciding the reactor configuration. The pressure drop between reactors is that required for the furnace coil. This together with total pressure is subject to further study involving furnace, recycle-compressor and catalyst costs. In making such a study various cases can be calculated and the optimum sought.* Engineering judgment must be applied with a strong hand, for the problem as stated lacks good quantitative data on activity

* Furnace coils of the type required in this unit can be designed for as low as 5 psi drop. This lower ΔP will reduce compressor costs and must be studied in comparison with increased furnace costs caused by the special coil design.

maintenance. Lower pressures improve the aromatics production rate but adversely affect activity. Because of the lack of precise expressions for activity, an elaborate optimization study is not indicated and, in fact, could lead to erroneous conclusions if the designer lost sight of the limited nature of the data. In fact a conservative decision in this particular situation might be to design the reactor and associated equipment for operating at the higher pressure range (575 psig), but purchase initially an amount of the expensive catalyst based on calculations for the lower pressure range (425 psig). The reactor design temperature shall, of course, be based on the end-of-run temperature (1000°F).

Reactor Configuration

Select a G such that ΔT between catalyst surface and bulk fluid is minimized, and at the same time ΔP for the bed is not excessive. It is clear based on preliminary calculations that the L/D will be low and great care in designing distributors and outlet collectors will be necessary.

Reactor 1

$$Z = \left(\frac{W}{F_T}G\right)\frac{1}{\rho_b M_F} = \frac{6.03\,G}{(48.7)(159)} = 7.79 \times 10^{-4}G, \qquad \text{ft}$$

$$D = \sqrt{\left(\frac{\text{total mass flow}}{G}\right)\frac{4}{\pi}} = \sqrt{\frac{329,335\,(4)}{G}\,\frac{}{\pi}} = 647.6(G)^{-0.5}, \qquad \text{ft}$$

Thus for $G = 3000\,\text{lb/hr ft}^2$, $Z = 2.34\,\text{ft}$ and $D = 11.8\,\text{ft}$. Using the procedures shown in Case Study 105 the corresponding $\Delta P = 2.44$ psi and the ΔT between catalyst and bulk fluid is 14°F (8°C) at the inlet, which is high. By increasing G to 6000 this is reduced to approximately 8°F (4.4°C), which is within accuracy of rate data. The corresponding dimensions will then be 8.36 ft in diameter and 4.67 ft of bed with a ΔP of 18.1 psi. This ΔP is less than 10% of the operating pressure and the larger bed height will reduce the problem of flow distribution. Final selection of bed dimensions for each reactor must be based on a more thorough study of costs including compression costs as in Case Study 105. Vessel costs may be lowered by specifying a refractory lining so that the design metal wall temperature may be lowered.

Adding additional catalyst to extend run length must be approached with caution in this complex reaction system. In addition to the disadvantage of higher catalyst cost with this expensive catalyst, more catalyst in the first bed will increase aromatics content of the effluent, which is desirable; but the paraffin content will increase further, which is undesirable.

Parametric Sensitivity

Several parameters were varied to ascertain the effect of possible errors on the final values of product quality. Selecting 943°F as a base temperature, activation energies were varied $\pm 10\%$ and the preexponential factor altered so that values of rate constants would remain the same at 943°F. These changes produced variations of only 1 % in volume % aromatics and yield. This simply illustrates that if the rate constant is "correct" at a temperature close to the average temperature where most of the conversion occurs (entrance of reactors), small errors in the activation energy do not effect the result for this system.

The model was found to be highly sensitive to the hydrogen content of the reaction mixture which in turn depends on the accuracy of the flash equilibrium at the separator where recycle gas is removed. A variation of hydrogen content of the recycle gas from 75.8 to 100 mole % produced a variation in volume % aromatics from 65.1 to 57.7%. For illustrative purposes this program used an arbitrary split at the separator. Ideally a flash calculation with precise equilibrium data together with a recycle material-balance routine should be included. The sensitivity of the model and the process to hydrogen pressure is caused by its strong effect on the aromatics reaction. Not included in the model, but also important, is the effect of hydrogen in reducing coking.

REFERENCES

1.   J. H. Sinfelt, *Adv. Chem. Eng.*, **5**, 38 (1964).
2.   R. B. Smith, *Chem. Eng. Prog.*, **55** (6), 76 (1959).
3.   W. H. Decker and C. Rylander, *Oil Gas J.*, **57** (6), 88 (1959).
4.   W. S. Kmak, *A Kinetic Simulation Model for the Power-forming Process*, A.I.Ch.E. National Meeting, Houston, Tex., March 3, 1971.

CASE STUDY 109

Ammonia Oxidation

THIS STUDY illustrates design aspects of a shallow-bed adiabatic reactor involving a very rapid reaction, which is diffusion controlled. The design of the bed, composed of a small number of catalyst screens, is normally based on past experience, scale-up on the basis of equal velocity, and careful planning for good distribution. A design model can be useful in economic or operating studies.

Problem Statement

Design a reactor for producing sufficient nitric oxide by partial oxidation of ammonia to supply a nitric acid plant having a capacity of 245 tons of nitric acid per day. Catalyst: 90% Pt + 10% Rh alloy as fine-mesh screen (p. 531[1]). Its average life is 3–9 mo. Depending on raw material, operating and catalyst costs, operating cycles vary from 30–90 days. After each cycle, one or more gauzes at the top is removed for reclaiming. The remaining gauzes are made the top portion and new gauzes added at the bottom of the stack.

Process Description

The direct oxidation of ammonia to nitric oxide (NO) over platinum catalyst is one of the major steps in the production of nitric acid. Subsequent steps include oxidation of NO to NO_2 and absorption of NO_2 in water to form HNO_3 (1,2). The direct oxidation step occurs in the range of 930°C and 100 – 125 psig. At this temperature platinum oxides are formed on the surface and a portion of these vaporize and are carried away in the gas stream (3).

The typical loss in a well-operated plant is 0.01–0.012 Troy oz/ton of nitric acid produced (3).

Operating conditions will, of course, alter these values. Poor distribution of gases that can induce localized hot spots in this exothermic system can materially increase the platinum loss. High plantinum costs give top priority to designs and operating procedures which minimize losses. Even though filter equipment for recovering platinum from the exit gases is normally specified, it is not 100% efficient. Between 30% and 50% of that volatilized is recoverable from the filter elements and by cleaning downstream equipment. To alleviate this problem further, a Pd/Au getter directly below the catalyst mesh is used (see Fig. 11.14, p. 531[1]) to capture volatilized platinum by alloying with it.

Chemistry

This very rapid reaction has been difficult to study mechanistically. The overall stoichiometric reaction $(4NH_3 + 5O_2 \rightarrow 4NO + 6H_2O)$ occurs by some complex mechanism. The following has been suggested.

$$O_2 \longrightarrow 2O$$
$$NH_3 + O \longrightarrow NH_2OH$$
$$NH_2OH \longrightarrow NH + H_2O$$
$$NH + O_2 \longrightarrow HNO_2$$
$$HNO_2 \longrightarrow NO + OH$$
$$2OH \longrightarrow H_2O + O$$

Typical of oxidation catalysts, it is thought that oxygen is chemisorbed on the catalyst. This chemisorbed oxygen then reacts with ammonia, producing a chemisorbed imide radical that is proposed here to react with molecular oxygen to yield finally nitric oxide.

Thermodynamics

The reaction is highly exothermic and not equilibrium limited. The following relations have been reported for this reaction system in the limited temperature range of interest (4–6).

$$(-\Delta H) = 54{,}250 - 0.4(T - 298) \text{ cal/g mole of } NH_3 \text{ reacted} \qquad \text{(CS-9.1)}$$

where $T = °K$

$$K = 2.11 \times 10^{19} @ 700°K$$

Basis: $NH_3 + \frac{5}{4}O_2 \rightarrow NO + \frac{3}{2}H_2O$

$$c_{p_m} = 6.45 + T(1.52 + 4.08y_A)10^{-3} \qquad \text{(CS-9.2)}$$

where y_A is the mole fraction of ammonia in feed. This heat capacity equation applies in the region of $T = 600–1100°K$ and $y_A = 0.09–0.12$.

Referring to Table 6.4 the adiabatic factor of 633 is large, but the heat generation potential is low because of the dilute system and the mass-transfer control.

For convenience the equations for viscosity and thermal conductivity of the mixture in the same temperature regions will also be given at this point (6).

$$\lambda_f = (3.31 + 12.77 \times 10^{-3}T)10^{-6} \frac{\text{k cal}}{\text{m sec}} \, °C \qquad \text{(CS-9.3)}$$

$$\mu_f = (12.5 + 29.20 \times 10^{-3}T)10^{-5} \text{ g/cm sec} \qquad \text{(CS-9.4)}$$

where $T = °K$.

Kinetics

Early experimental work on this process demonstrated that under normal operating conditions the physical transport of ammonia to the catalyst surface is the rate determining step (7). At low temperatures or with deactivated catalyst the chemical rate can become limiting or partially limiting. With active catalyst, however, it is clear that high temperatures and high velocities will produce essentially total conversion of ammonia. The maximum temperature is limited by the aforementioned catalyst loss that becomes excessive much above 900°C. The loss per unit surface area is apparently a direct function of gauze temperature, mass flow rate, and ammonia-to-oxygen ratio in the feed. The temperature dependency in the region 825 to 900°C is steep and expressed (4) approximately as $10^{-20000/T}$.

On the basis of mass-transfer control the rate of ammonia oxidation may be written in terms of a mass-transfer coefficient with the ammonia partial pressure at the catalyst surface assumed to be zero for this rapid reaction.

$$(-r_A) = k_{g_A}{}^s a_{WR} P_A \qquad \text{(CS-9.5)}$$

where P_A is the partial pressure of ammonia in bulk fluid and a_{WR} is the surface area per unit volume of screen. For shallow beds of the type shown on p. 531[1], axial diffusion must surely be important, but the very high flow rates encountered in commercial equipment cause it to be less important than the problem of equal flow distribution across the gauzes. By including a gas distributor below the catalyst pad, as described on p. 531[1], this problem is greatly alleviated, and as a first approximation a model is developed based on plug flow.

$$\frac{-G}{M_F} dy_A = (-r_A)dZ \qquad \text{(CS-9.6)}$$

where $(-r_A)$ is the g-moles of NH_3 converted/cm^{3}sec.

For this rapid reaction assume that partial pressure of ammonia on the catalyst surface is zero. Then the rate of transfer of ammonia becomes (6)

$$(-r_A)dZ = k_{g_A}{}^s a_{WR} y_A P dZ = k_{g_A}{}^s y_A P f_w dn_s \qquad \text{(CS-9.7)}$$

where n_s is the number of gauzes and f_w is the wire area per gauze cross-sectional area for one gauze.

$$\frac{-G}{M_F} dy_A = k_{g_A}{}^s y_A P f_w dn_s \qquad \text{(CS-9.8)}$$

or integrating and simplifying by assuming constant number of moles

$$k_{g_A}{}^s P n_s f_w = \frac{G}{M_F} \ln \frac{y_{A_1}}{y_{A_2}} = \frac{G}{M_F} \ln \frac{y_{A_1}}{y_{A_1}(1 - X_A)} = \frac{G}{M_F} \ln \frac{1}{1 - X_A}$$

or

$$\ln(1 - X_A) = -k_{g_A}{}^s P n_s f_w \frac{M_F}{G} \qquad \text{(CS-9.9)}$$

The following substitutions, which conveniently eliminate some variables follow the procedure of Ref. 6, except more recent mass-transfer data are used (8). From p. 532[1],

$$k_{g_A}{}^s = \frac{0.865 N_{Re}{}^{-0.648} G}{P N_{Sc}{}^{\frac{2}{3}} M_m \varepsilon_w}$$

where M_m is the molecular weight of mixture.

$$\ln(1 - X_A) = \frac{-0.865 n_s f_w M_F N_{Re}{}^{-0.648}}{\varepsilon_w M_m N_{Sc}{}^{\frac{2}{3}}}$$

$$= \frac{-0.865 n_s f_w \varepsilon^{0.648} \mu_f{}^{0.648} \mathscr{D}^{\frac{2}{3}} \rho_f{}^{\frac{2}{3}}}{\varepsilon_w \mu_f{}^{\frac{2}{3}} d_w{}^{0.648} G^{0.648}} \qquad \text{(CS-9.10)}$$

where d_w is the diameter of wire. M_F/M_m is approximately unity.

$$\ln(1 - X_A) = - \frac{(0.865) n_s f_w \mathscr{D}^{\frac{2}{3}} \rho_f{}^{\frac{2}{3}}}{\varepsilon_w{}^{0.352} d_w{}^{0.648} G^{0.648} \mu_f{}^{0.019}} \qquad \text{(CS-9.11)}$$

Using

$$\mathscr{D} = 0.227 \left(\frac{T}{293}\right)^{\frac{3}{2}} \left(\frac{1}{P}\right)^{*} \text{cm}^2/\text{sec}$$

$$\rho_f = \frac{MP}{RT} \text{ and } \mu_f \text{ from Eq. CS-9.4}$$

$$M = M_F = (32)(0.21)(1 - y_A)_0 + (28.01)(0.79)(1 - y_A)_0 + 17.03 y_{A_0}$$
$$= 28.85 + 11.82 y_{A_0}$$

$$\ln(1 - X_A) = -\frac{5.81761 \times 10^{-5} n_s f_w T_i^{0.333}(28.85 + 11.82 y_{A_0})^{0.667}}{\varepsilon_w^{0.352} d_W^{0.648} G^{0.648} \mu_f^{0.019}}$$

$$(\text{CS-9.12})$$

Since $\mathscr{D}$ has been introduced as cm^2/sec and R as cm^3 atm/g-moles $^\circ$K, $d_w = \text{cm}$, $G = \text{g/cm}^2\text{sec}$, and $\mu_f = \text{g/cm}$ sec. T_i is the mean film temperature.

The temperature dependency in Eq. CS-9.12 is relatively minor. Heat-transfer calculations based on assuming a constant gauze temperature corresponding approximately to the outlet gas temperature indicate a mean film temperature of 50° less than the exit temperature. Thus it is within the present accuracy of transport properties to simply use the exit gas temperature as the film temperature (12). This is the commonly measured temperature in commercial units. The inlet gas temperature may be calculated from an adiabatic heat balance with an exit gas temperature 20°C less than theoretical to account for incomplete combustion and radiation losses (4).

When these equations are applied to conditions for which experimental data are known, the calculated conversions are higher than the experimental values. It has been shown recently that in high-pressure plants with up to 50 gauze pads only the upper $\frac{2}{3}$'s of the bed appears to be active as a catalyst, and the lower $\frac{1}{3}$ is not (3). But when the extra screens are removed, the conversion declines because the reduced ΔP causes uneven distribution and bypassing of portions of the catalyst. Thus Eq. CS-9.12 must be multiplied by a correction factor (6). Experimental data involving small numbers of screens (≈ 3) agree when the right-hand side of Eq. CS-9.12 is multiplied by two-thirds (6), which fortuitously corresponds to the aforementioned $\frac{2}{3}$'s of the bed (3).

Design

This reaction system is another example of an old process for which catalyst suppliers provide special catalyst and are helpful in recommending the

---

* $\mathscr{D} = 0.227 \text{ cm}^2/\text{sec}$ @ 293°K for NH_3 in air at 1 atm (9).

configuration and amount of catalyst based on knowledge of many operating plants. In such cases a model for the reactor is not as important in initial design as it is in subsequent studies concerning economics associated with catalyst replacement, operating strategy and yield.

Typical reported requirements at 900°C gauze temperature and 100 psig are a quantity of 80 mesh gauze with wire of 0.003 in. in diameter equivalent to 2 Troy oz/ton of acid produced and a cross-sectional area of 2.7547 sq ft/100 daily tons of HNO_3 produced (3). A conversion of 96% would be expected at the outset (3). The ammonia would be adjusted to be 10% of the total feed. The explosive limit is reported to be 13% (4) and the usual operating range is between 8 and 11.5%.

Equal performance is obtained with 40–45% of this amount of catalyst if a gas distributor is used, as described on p. 513[1] (3). Thus the desired 245-ton plant for this alternate would be specified to have 0.8–0.9 Troy oz of Pt-Rh/100 ton acid, two to four sheets of Degussa getter for Pt recovery, and the aforementioned gas distributor. The cross-sectional area and mass velocity for this 245-ton plant are:

Cross Sectional Area

$$2.7547 \text{ sq ft}/100 \text{ tons } HNO_3$$

$$(2.7547)(2.45) = 6.749 \text{ sq ft or } 33.5 \text{ in.}$$

hexagonal gauze (19.34 in. side)

Weight of 80 mesh Pt-Rh gauze $= 1.71$ Troy oz/ft^2. Use 2 Troy oz/ton HNO_3.
$\therefore$ Number of gauzes $= [(2)(245)]/[(6.749)(1.71)] = 42.46$ or 43 gauzes. Alternatively, a gas distributor and Degussa getter would improve distribution and reduce the number of gauzes. Use the higher value 0.9 Troy oz/ton as conservative value.

$$\left(\frac{0.9}{2}\right)(42.46) = 19.1 \text{ gauzes}$$

Inlet and outlet distributions, as shown in Fig. 11.14, should be designed in accordance with the criteria set forth on p. 309[1] in order to maximize pressure recovery and assure good distribution. A magnetic separator should be located in the liquid ammonia stream to remove iron oxide, which accelerates gauze deterioration (10).

Ideally if an elbow must be used, it should be at least 10 diameters upstream from the diffuser. If less, circular arc vanes must be installed in the elbow, as discussed on p. 309[1]. To minimize diffuser cost, it may be necessary to use larger angles than indicated on p. 309[1]. In such cases multiple screen distributors should be installed downstream of the throat. Alternatively, extrapolating Fig. 7.3, a 30° diffuser may be selected with a 12-in. diameter inlet pipe. The outlet diffuser would be similar. It is most important in this case to avoid a close elbow. Many units have been carefully designed at the diffuser and then installed immediately after the elbow thereby negating the effect of the diffuser. Further experimental work should be done on the optimum arrangement for such short-bed reactors.

Design with Model

It is now instructive to determine the usefulness of the proposed model by comparing predicted performance with the specification of 19 gauzes when a gas distributor is used.

Basis: 100 psig and 900°C gauze temperature; 10% NH_3 in feed. Gauze mesh size 80 in.$^{-1}$ and diameter 0.003 in. From Ref. 28, Chapter 7 and based on plain square weave with a thickness of $2d_w$, the surface area per unit volume, a_{WR}, for the gauze and other properties are:

$$a_{WR} = \pi l_w n_w{}^2$$

where $l_w = [(1/n_w)^2 + d_w{}^2]^{0.5}$

$$a_{WR} = \pi n_w{}^2 \sqrt{\left(\frac{1}{n_w}\right)^2 + d_w{}^2} = \pi(80)^2 \sqrt{\frac{1}{(80)^2} + 0.003^2}$$

$$= 258 \text{ in.}^{-1} \text{ or } 101.7 \text{ cm}^{-1}$$

where n_w is the mesh size and d_w is the wire diameter.

$$f_w = a_{WR} 2d_w = (258)(2)(0.003) = 1.55$$

$$\varepsilon_w = 1 - \frac{\pi l_w n_w{}^2 d_w}{4} = 1 - \frac{a_{WR} d_w}{4} = 0.806$$

Mass Velocity

$$G = \frac{(1)(2000)(30.03/63)}{(24)(3600)(0.96)(0.1)(0.027547)} = 4.17 \text{ lb/ft}^2 \text{ sec or } 2.04 \text{ g/cm}^2 \text{ sec}$$

This corresponds to a superficial velocity based on outlet conditions of

$$u_s = (2.04)\frac{1}{\rho_f} = (2.04)\frac{(82.06)(1203)}{(115/14.7)(30.03)} = 857 \text{ cm/sec or } 28 \text{ ft/sec}$$

Thus $u_s \approx 28$ ft/sec constitutes a more general criterion based on these reported data.

Equation CS-9.12 provides a means for estimating the number of gauzes when perfect plug flow is assumed which is approached in the unit with a special gas distributor.

$$n_s = \frac{-\ln(1 - X_A)\varepsilon^{0.352}d_w^{0.648}G^{0.648}\mu_f^{0.019}}{(5.81761 \times 10^{-5})f_w T_i^{0.333}(28.85 + 11.82y_{A_0})^{0.667}}$$

$$n_s = \frac{(-\ln 0.04)(0.806)^{0.352}(0.0076)^{0.648}(2.04)^{0.648}(41.99 \times 10^{-5})^{0.019}}{(5.81761 \times 10^{-5})(1.55)(1173)^{0.333}(30.03)^{0.667}}$$

$$= 18.8$$

Since the predicted mass-transfer coefficient has a mean deviation of $\pm 12\%$, the number of gauzes is well within experimental error. A more useful model would include the decline in conversion with time which is caused by deactivation. The apparent mass-transfer coefficient is reduced by this phenomenon as the resistance caused by the chemical reaction becomes finite. If this effect can be correlated with temperature and time, various operating strategies can be studied. Higher temperatures will increase yield but also platinum losses. A complete economic assessment would be possible if such predictive tools were known.

Equation CS-9.11 can be useful in comparing various gauze configurations with performance using efficiency factors observed for operating plants. More recent mass-transfer data have been reported (11) and when substituted into Eq. CS-9.9, approximately the same result is obtained (see also Ref. 12).

REFERENCES

1. T. H. Chilton, *Strong Water*, MIT Press, Cambridge, Mass., 1968.
2. T. H. Chilton, *Chem. Eng. Prog. Monograph Ser.*, **56** (3), (1960).
3. G. R. Gillespie and R. E. Kenson, *Chem. Tech.*, p. 627 (Oct., 1971).
4. A. P. Delle, *Chem. Eng. Sci.*, **8**, 146 (1958).
5. R. H. Harrison and K. A. Kobe, *Chem. Eng. Prog.*, **49**, 349 (1953).
6. I. Curievici and St. Ungureanu, *Bul. Inst. Poli Din. Iasi*, **14**, 227 (1968).
7. L. Andpussow, *Z. angew. Chem.*, **32**, 321 (1926) and **40**, 166 (1927).
8. C. N. Satterfield and D. H. Cortez, *Ind. Eng. Chem. Fundam.*, **9**, 613 (1970).
9. S. P. S. Andrew, *Chem. Eng. Sci.*, **4**, 269 (1955).
10. G. R. Gillespie and D. Goodfellow, *Chem. Eng. Prog.*, **70** (3), 81 (1974).
11. M. A. Shah and D. Roberts, *Adv. Chem. Ser.*, **133**, 259 (1974).
12. D. Roberts and G. R. Gillespie, *Adv. Chem. Ser.*, **133**, 600 (1974).

CASE STUDY 110

Phthalic Anhydride Production

THIS is a classic example of a rapid oxidation reaction which if not controlled, produces CO and CO_2 rather than the desired product. A multitubular reactor is essential and techniques for obtaining nearly complete conversion without causing a runaway must be developed.

The kinetic data used as representative of bench-scale studies have been presented previously for modeling studies as a means for discussing reactor models (4). The source and limitations of the data were not discussed. These data, however, are useful for this study since they have been employed previously in model studies and the present case study can add to the background already so expertly presented (4).

Problem Statement

An existing plant for producing phthalic anhydride from naphthalene is to be converted to o-xylene feed. The reactor consists of 2500 tubes, 2.5 cm ID and with a packed catalyst bed height of 3 m. The heat-transfer medium circulating outside the tubes is a sodium nitrite-potassium nitrate, fused-salt (HTS, heat-transfer salt). The elevated temperatures (350–400°C) required for these reactions make the use of boiling water impractical because of the high shell-side pressures that would be required. Steam is generated, however, in water cooling coils that are located in the fused-salt bath. The fused salt is continuously agitated by a mechanical stirrer (5). During the days of Houdry fixed-bed cracking of petroleum much experience was gained with heat-transfer salt, and it has been documented (6,7).

A pilot plant, consisting of a molten-salt bath containing one tube of the same dimensions as those in the large reactor, has been constructed and a new catalyst has been studied briefly using o-xylene as the feed. At a mass velocity, $G = 4684$ kg/m^{2}hr necessary to maintain the desired production of phthalic anhydride in existing reactors at complete conversion of xylene and maximum selectivity, it was found impossible with the new catalyst to reach a conversion over 67% in the 3-m length at 360°C. Raising this temperature any significant amount caused reaction runaway.

Some preliminary bench-scale studies with the new catalyst have yielded some rate data that can be used for modeling this system.

Develop a model and test it at the pilot-plant conditions. Consider means for overcoming the production problem with the new catalyst and test these with the model. Ideas which are indicated to be valid by calculation and which, of course, make physical sense can then be confirmed on the pilot plant.

Chemistry

Overall Reactions

$$C_8H_{10} + 3O_2 \longrightarrow C_8H_4O_3 + 3H_2O;$$

$$C_8H_{10} + 10\tfrac{1}{2}O_2 \longrightarrow 8CO_2 + 5H_2O)$$

The oxidation of o-xylene occurs on V_2O_5 catalyst with good selectivity for phthalic anhydride, provided temperature is controlled within relatively narrow limits. To avoid explosive mixtures the o-xylene content of the feed stream of air and o-xylene should not exceed one percent.

The catalyst is thought to change oxidation states during the reaction such that the reduced form adds oxygen which in turn oxidizes the hydrocarbon, returning the catalyst to the reduced state (1). The total reaction scheme is quite complex and involves series and parallel reactions, as suggested in Fig. CS-10.1 (3). Side reactions to maleic anhydride and benzoic acid, which are not shown, produce the major byproducts other than CO and CO_2. The intermediates TA and PI are not present in the product at high conversions of o-xylene.

Catalyst Properties

Most catalysts for o-xylene and naphthalene oxidation contain V_2O_5 as the active component. Various improved catalysts have been introduced which give longer life and better selectivity. One type with good selectivity employs a low area inert support such as α-alumina with a bulk density of 1.3 g/cm^3. It consists of 0.3-cm spheres and has a 10% V_2O_5 content.

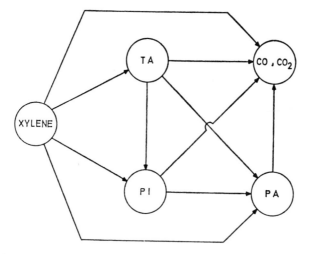

Fig. CS-10.1 Phthalic anhydride reaction scheme (TA = o-tolualdehyde, PI = phthalide, PA = phthalic anhydride). Reproduced by permission: J. Herten, and G. F. Froment, *Ind. Eng. Chem. Process Des. Develop.*, **7**, 517 (1968). [Copyright by the American Chemical Society.]

Thermodynamics

The reactions are highly exothermic. From Table 6.4 the high adiabatic factors (378 and 1344) and heat-generation potentials (12.8 and 48.3) confirm the need for heat-transfer reactors. Careful control is necessary to avoid ranges of temperature where total oxidation becomes dominate because of the extreme exothermicity of this reaction. A fluidized-bed reactor with internal heat-transfer surface is an alternate to the fixed-bed, multitubular reactor, and both have been successfully employed.

Safety

Feed composition must not exceed 1 mole % o-xylene in air so that an explosive mixture is avoided.

Kinetics

The oxidation of organic compounds such as o-xylene is highly complex. Models for the disappearance of o-xylene have been presented, one involving a steady-state oxidation–reduction assumption (1) and the other a steady state with equality of rate of oxygen adsorption and rate of reaction (2).

Both of these yield the same form of rate equation. As is often the case, the fit of experimental data does not resolve a mechanistic search.

For practical purposes more than just the rate of disappearance of o-xylene is needed. In order to model the highly temperature sensitive system at the very least the competition between PA production and complete oxidation must be followed. The complete oxidation causes extremely large heat effects which cannot be ignored.

It is necessary, as in all such complex systems, to simplify the structure of Fig. CS-10.1 to the following (4):

$$o\text{-xylene} \xrightarrow{\text{(air)}k_1} \text{PA} \xrightarrow{\text{(air)}k_2} CO_2 + CO$$
$$\xrightarrow{\text{(air)}k_3} CO_2 + CO$$

The corresponding rate expressions and rate constants are (4):

$$\hat{r}_1 = \hat{k}_1 P_A P_O; \ln k_1 = \frac{-27{,}000}{R'T} + 19.837 \qquad \text{(CS-10.1)}$$

$$\hat{r}_2 = \hat{k}_2 P_B P_O; \ln k_2 = \frac{-31{,}000}{R'T} + 20.86 \qquad \text{(CS-10.2)}$$

$$\hat{r}_3 = \hat{k}_3 P_A P_O; \ln k_3 = \frac{-28{,}600}{R'T} + 18.97 \qquad \text{(CS-10.3)}$$

where A is the o-xylene, B is phthalic anhydride, O is the oxygen, and rates are in terms of kg moles converted/(kg cat.)(hr).

Equations CS-10.1–CS-10.3 can be rewritten in terms of conversion (4).

Basis: 1 mole o-xylene and constant mole fraction of oxygen (O_2 is present in large excess)

$$\hat{r}_B = \hat{r}_1 - \hat{r}_2 = P^2 y_{O_0} y_{A_0} [k_1 (1 - X_B - X_C) - k_2 X_B] \qquad \text{(CS-10.4)}$$

$$\hat{r}_C = \hat{r}_2 + \hat{r}_3 = P^2 y_{O_0} y_{A_0} [k_2 X_B + k_3 (1 - X_B - X_C)] \qquad \text{(CS-10.5)}$$

where X_B is net moles PA (B) formed per mole o-xylene (A) charged, X_C is moles xylene converted to CO and CO_2 per mole o-xylene (A) charged.

Design Model for Heat-Transfer Reactor

For illustrative purposes both a one-dimensional and two-dimensional model will be used. The strategy illustrated here will be to make preliminary calculations with a one-dimensional model followed by estimates of the probable uncertainties due to interphase resistances and radial gradients.

Whatever these results show, a two-dimensional model will also be used to illustrate its application and to compare the results directly. In addition the algorithm developed here for the two-dimensional model will be compared with the published study which used the same data but a different algorithm.

One-Dimensional Model

Equations 9.3B and 9.4A of Table 9.1 for plug flow and negligible interphase gradients can be written in terms of the appearance of PA and CO + CO$_2$ with an inerts term added if the bed is diluted with inert support.

$$\frac{dX_B}{dZ} = \frac{M_F \hat{r}_B \rho_b}{y_{A_0} G(1 + b_I)} \tag{CS-10.6}$$

$$\frac{dX_C}{dZ} = \frac{M_F \hat{r}_C \rho_b}{y_{A_0} G(1 + b_I)} \tag{CS-10.7}$$

$$\frac{dT}{dZ} = -\frac{4U}{DGc_p}(T - T_j) + \frac{\rho_b\left(\dfrac{\hat{r}_B}{1 + b_I}\right)(-\Delta H_B)}{Gc_p} + \frac{\rho_b\left(\dfrac{\hat{r}_C}{1 + b_I}\right)(-\Delta H_C)}{Gc_p} \tag{CS-10.8}$$

where b_I is the mass of inerts per unit mass of catalyst.

The high heat-transfer coefficients obtainable with molten salts allows one to neglect the shell-side resistance along with the metal tube wall in comparison with the internal heat-transfer resistance.
Hence $h_T = h_w$.
From Eq. 11.53

$$\frac{1}{U} = \frac{1}{h_w} + \frac{D}{8\lambda_r} \tag{CS-10.9}$$

The values of h_w and λ_r given in Ref. 4 will be used so that the calculations can be compared. These values are slightly different from those that would be determined from the more recent correlation given in Appendix D. At $G = 4684 \text{ kg/m}^2\text{hr}$, $h_w = 134 \text{ kcal/m}^2\text{hr}°\text{C}$, and $\lambda_r = 0.67 \text{ kcal/mhr}°\text{C}$ from which $U = 82.7 \text{ kcal/m}^2\text{hr}°\text{C}$. Other required data are $\Delta H_B = -307 \text{ kcal/g}$ mole xylene to PA, $\Delta H_C = -1,090 \text{ kcal/g}$ mole xylene to CO and CO$_2$, $y_{A_0} = 0.0093$, $y_{O_0} = 0.208$, $\rho_b = 1300 \text{ kg/m}^3$, $D_p = 0.003$ m, $D = 0.025$ m, $c_p = 0.25 \text{ kcal/kg}°\text{C}$ and $M_F = 29.48$.

Equations CS-10.6–CS-10.8 were solved using a conventional difference scheme. Values of R_q (Eq. 11.58) were calculated at each increment. The average rate for the first increment was estimated from $(r_B)_1 = (r_B)_0 + 0.04$

ΔZ and $(r_C)_1 = (r_C)_0 + 0.07 \, \Delta Z$. At other increments $\bar{r} = (r_{n-2} - r_{n-1})(\frac{1}{2}) + r_{n-1}$. The ΔT, ΔX, and ΔW for the increment were calculated and then the average rate recalculated and an iteration on temperature, conversion, and rate followed until convergence occurred.

Because of the long life of V_2O_5 catalyst, it is reasonable to consider diluting the bed with inerts or employing a less active form in the early portion of the bed in order to avoid excessive hot spots. Both of these techniques are included in existing patents on specific catalysts for this process (9,10). In employing either of these strategies the goal is to increase inlet and cooling temperatures so that complete conversion can be realized. The higher activation energy for the side reactions, however, definitely limits the extent of this increase.

Results with One-Dimensional Model

The results of several strategies are summarized in Table CS-10.1 and Fig. CS-10.2. With a value of $q_p = 48.3$ as maximum and values of R_q exceeding unity only near temperature maxima and then not greatly, the one-dimensional model should be reasonably valid. This is especially true for those cases removed from runaway (see p. 543[l]).

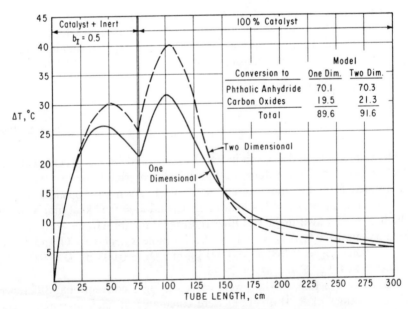

Fig. CS-10.2 Temperature profiles. (Alternate 6.)

Table CS-10.1 Alternate Designs for Phthalic-Anhydride Reactor[a]

Alternate Number	Inert Distribution	Inlet Temp. °C	Conversion[b] to Phthalic Anhydride	Conversion[b] to CO and CO_2	Total Conversion	% Selectivity[c]	Yield Lb PA per 100 Lb o-xylene
1	0	357	0.6377	0.1277	0.7654	83.32	89.04
2	0	360	0.6692	0.1486	0.8178	81.83	93.44
3	0	365	Runaway				
4	$b_1 = 0.5$ in first $\frac{1}{2}$ $b_1 = 0$ in last $\frac{1}{2}$	375	0.7017	0.2101	0.9118	76.96	97.97
5	$b_1 = 0.5$ for 1st $\frac{1}{4}$ $b_1 = 0$ in last $\frac{3}{4}$	375	0.7016	0.2382	0.9398	74.65	97.96
6	$b_1 = 0.5$ for 1st $\frac{1}{4}$ $b_1 = 0$ in last $\frac{3}{4}$	370	0.7010	0.1949	0.8959	78.25	97.86
7	$b_1 = 0.5$ for first $\frac{1}{2}$ $b_1 = 0$ for last $\frac{1}{2}$	380	Runaway				

[a] Reactor tube length is 3 m.
[b] Moles o-xylene converted to indicated product per mole of xylene charged.
[c] Selectivity = (conversion to PA)/(total conversion).

129

Alternate 6 in Table CS-10.1 is recommended as the best course of action involving charging $\frac{3}{4}$'s of the downflow reactor length with catalyst and the remaining top $\frac{1}{4}$ with a 50–50 mixture of catalyst and inert support. This causes two smaller hot spots instead of one larger uncontrollable temperature surge and a high enough temperature may thus be used to assure high conversion of o-xylene. The residual xylene may be consumed in a second reactor operated to ensure conversion rather than selectivity, or the residual gases may be used as an energy source.

Two-Dimensional Model

Since the one-dimensional model requires far less computing time, it is preferred for exploratory work. It is always feasible to check the crucial decisions thus reached using a two-dimensional model for comparison.

Equations 9.20 and 9.21 can be applied to this system with boundary conditions, as given on p. 408[1].

$$\frac{\partial X_B}{\partial Z} = \frac{\mathscr{D}_r \rho_f}{G} \left(\frac{\partial^2 X_B}{\partial \mathbf{r}^2} + \frac{1}{\mathbf{r}} \frac{\partial X_B}{\partial \mathbf{r}} \right) + \frac{\hat{r}_B \rho_B M_F}{G y_{A_0}} \qquad \text{(CS-10.10)}$$

$$\frac{\partial X_C}{\partial Z} = \frac{\mathscr{D}_r \rho_f}{G} \left(\frac{\partial^2 X_C}{\partial \mathbf{r}^2} + \frac{1}{\mathbf{r}} \frac{\partial X_C}{\partial \mathbf{r}} \right) + \frac{\hat{r}_C \rho_B M_F}{G y_{A_0}} \qquad \text{(CS-10.11)}$$

$$\frac{\partial T}{\partial Z} = \frac{\lambda_r}{G c_p} \left(\frac{\partial^2 T}{\partial \mathbf{r}^2} + \frac{1}{\mathbf{r}} \frac{\partial T}{\partial \mathbf{r}} \right) + \frac{\rho_B(-\Delta H_B)}{G c_p} \hat{r}_B + \frac{\rho_B(-\Delta H_C)}{G c_p} \hat{r}_C \qquad \text{(CS-10.12)}$$

The rate terms should be divided by $1 + b_1$ when inert solids are used.

The difference equations were derived by, as earlier suggested (11), transforming the radial coordinate of the differential equations which all have the same dimensionless form. The complex derivation yielded the following equations (12).

General Dimensionless Equation

$$\frac{\partial y}{\partial Z_a} = a_i \left[\frac{1}{\mathbf{r}_a} \frac{\partial y}{\partial \mathbf{r}_a} + \frac{\partial y}{\partial \mathbf{r}_a} \right] + S_i$$

where y refers to variable X_B, X_C, or $T_a = T - T_j$; $a_i = \mathscr{D}_r \rho_f / G D_p$ and $\lambda_r / G c_p D_p$ for heat balance equation; $\Delta Z_a = \Delta Z / D_p$, $\mathbf{r}_a = \mathbf{r}/D_p$, and $S_i = b_1 \hat{r}_B$ or $b_1 \hat{r}_C$ for mole balance and

$$\left[\frac{\rho_b D_p(-\Delta H_B)}{G c_p} \right] \hat{r}_B + \left[\frac{\rho_B D_p(-\Delta H_C)}{G c_p} \right] \hat{r}_C$$

for heat balance, and $b_1 = \rho_b D_p M_m / G y_{A_0}$. For interior of bed ($1 < n < N_E$, where N_E is the number of equal area radial increments):

$$y_{n,L+1} = y_{n,L} + \frac{2\Delta Z_a a_i}{\Delta u} [(2n - 1)y_{n+1,L} + (2n - 3)y_{n-1,L}$$

$$- (4n - 4)y_{n,L}] + \Delta Z_a S_i$$

where L and n are the number of axial and radial increments, respectively, and $u = \mathbf{r}_a^2$. For $n = 1$

$$y_{1,L+1} = y_{1,L} + \frac{2\Delta Z_a a_i}{\Delta u} (-3y_{1,L} + 4y_{2,L} - y_{3,L}) + \Delta Z_a S_i$$

For $n = N_E$:

$$y_{N_E,L+1} = y_{N_E,L} + \frac{8a_i \Delta Z_a}{\Delta u} \left[\frac{Dh_w T_a}{4\lambda_r} \delta_{T_a,y} - (N_E - \tfrac{3}{2})(y_{N_E,L} - y_{N_E-1,L}) \right]$$

$$+ \Delta Z_a S_i$$

where $\delta_{T_a,y} = 0$ if $y \neq T_a$ and $= 1$ if $y = T_a$.

Results with Two-Dimensional Model

The algorithm for the two-dimensional model is similar to that described for the one-dimensional model. The difference equation routine was used to solve for T_a, X_B and X_C at each Z_a and r_a position. Stability was assured by maintaining $\Delta Z_a \leq \tfrac{1}{2}\Delta \mathbf{r}_a^2$. The equal area method with a six radial increments gives good agreement with the previously reported Crank-Nicholson method (4,8) while equal radial steps approach these solutions only at an impractical number of steps. The computing time using six equal-area increments was 5.68 sec for a 3-m tube.

The average radial temperature at various axial positions is plotted for alternate 6 in Fig. CS-10.2. Although these profiles do not exactly correspond, they are close enough to warrant using the one-dimensional model as the criteria given on p. 543[1] and applied on p. 128 already suggested. The slightly higher conversion to carbon oxides for the two-dimensional model is caused by the higher temperature excursion.

REFERENCES

1. P. Mars and D. W. van Krevelen, *Chem. Eng. Sci.*, **3**, 41 (1954).
2. J. A. Juusola, R. F. Mann, and J. Downie, *J. Catal.*, **17**, 106 (1970).
3. J. Herten and G. F. Froment, *Ind. Eng. Chem. Process. Des. & Develop.*, **7**, 516 (1968).

4.   G. F. Froment, *Ind. Eng. Chem.*, **59**, 21 (1967).
5.   E. Guccione, *Chem. Eng. (N.Y.)*, p. 132 (June 7, 1965).
6.   W. B. Johnson and W. M. Nagle, *Ind. Eng. Chem.*, **39**, 971 (1947).
7.   R. H. Newton and H. G. Shimp, *Trans. A.I.Ch.E.*, **41**, 197 (1945).
8.   G. C. Grosjean and G. F. Froment, *Ned. Kon. VI Acad. Belg.*, **24**, 1 (1962).
9.   W. Friedrichsen and O. Goehre, (Badische Anilin u. Soda-Fabrik, A.G.) German Patent 2020482, Nov. 11, 1971.
10.   W. R. Grace & Co. Brit. Patent 1280703, July 5, 1972.
11.   J. Beek, *Adv. Chem. Eng.*, **3** (1962).
12.   G. R. Dowling, private communication, 1972.

CASE STUDY 111

Steam Reforming

THIS IMPORTANT PROCESS, the function of which is to generate hydrogen for such uses as ammonia synthesis (Fig. CS-5.1), affords an opportunity to demonstrate equilibrium and equilibrium-approach techniques in design (see p. 549[1]). Operating conditions are purposely selected to correspond to a reported industrial installation so that the calculated results may be compared with the known operating data (1).

Because of the high endothermic heat of reaction and the rapidity of the reactions, high heat fluxes, attainable conveniently with a direct-fired furnace, are required.

Problem Statement

Determine the number of 5-in. ID × 40 ft tubes required for a methane steam reforming unit to produce 50 million SCFD of H_2 with 95% purity when dry (60°F and 1 atm).

Feed Composition	Mole %
H_2O	84.07
H_2	1.56
CH_4	12.83
C_2H_6	0.61
C_3H_8	0.27
C_4H_{10}	0.07
N_2	0.58

Steam-to-carbon ratio: 5.5528

Catalyst

The catalyst is in the form of $\frac{5}{8}$ in. $\times$ $\frac{5}{8}$ in. $\times$ $\frac{3}{16}$ in. Raschig rings with $\rho_b =$ 53 $\pm$ 4 lb/ft^3, crush strength 90 lb, fusion temperature of 2500°F, and negligible shrinkage up to 2300°F. It contains 15% Ni and a maximum of 0.03% S and 0.2% SiO$_2$. Poisons are as listed in Table 2.13. Catalyst life varies from 1 to 5 yr depending on treatment. Excessive temperatures above 1000°C for periods longer than 1 hr cause irreversible growth of nickel crystallites and corresponding loss in active surface area (5). At low steam rates carbon deposits rapidly and causes the catalyst to disintegrate.

Chemistry and Thermodynamics

It is generally conceded that the following reactions can occur (1,2,4–6).

$$CH_4 + H_2O \longrightarrow 3H_2 + CO \tag{1}$$

$$CH_4 + 2H_2O \longrightarrow 4H_2 + CO_2 \tag{2}$$

$$CO + H_2O \longrightarrow CO_2 + H_2 \tag{3}$$

$$CH_4 + CO_2 \longrightarrow 2CO + 2H_2 \tag{4}$$

$$CO + H_2 \rightleftharpoons C + H_2O \tag{5}$$

$$CH_4 \rightleftharpoons C + 2H_2 \tag{6}$$

$$2CO \rightleftharpoons C + CO_2 \tag{7}$$

This group of reactions does not represent a mechanism, but is consistent with the known chemistry. Any pair among the first four reactions is adequate for representing equilibrium compositions, and it has been customary to select reactions 1 and 3. This conclusion applies when the ratio of steam to methane is high enough to prevent the presence of carbon at equilibrium (8). For carbon not to be present it is necessary that the steam to methane ratio be such that the following apply as calculated from simultaneous equilibrium of reactions 1 and 3.

$$\frac{a_{H_2O}}{a_{H_2} a_{CO}} \geq K_5, \quad \frac{a_{H_2}^2}{a_{CH_4}} \geq K_6, \quad \text{and} \quad \frac{a_{CO_2}}{a_{CO}^2} \geq K_7 \quad \text{(CS-11.1)}$$

Under these conditions any carbon formed will consume product and form reactants in the respective reactions until the value of K is reached. It has been shown that as the ratio applying to reaction 7 is satisfied, the other two are also satisfied (5,8). Thus the minimum steam ratio (moles steam-to-methane in feed) corresponds to $a_{CO_2}/a_{CO}^2 = K_7$. This theoretical value is invariably well below the ratios normally employed in modern reactors.

The combined or overall reaction comprising reactions 1 and 3 (CH_4 + $2H_3O \rightarrow CO_2 + 4H_2$) has an endothermic heat of reaction of 79,900 BTU/lb mole at $1000°F$ and increases approximately 650 BTU for every $100°F$ increase in reaction temperature. The very negative adiabatic factor (-682) and $q_p = -24.5$ (Table 6.4) confirm the need for a direct-fired furnace, as shown in Fig. 11.17 (p. 551[1]).

Since the gas as received for reforming is under pressure, energy is saved by operating the reformer at elevated pressures even though an increase in moles is involved. The negative effects of this higher pressure on the forward progress of reaction 1 can be overcome by increased temperature and high steam-to-methane ratios. The operating range to accomplish this favorable conversion requires high radiant flue-gas temperatures, and the combustion gases leaving the furnace contain a great deal of useable energy (9). A major portion of the furnace design is thus concerned with heat recovery usually by generating high-pressure steam.

Although kinetics and mechanism have been studied and it is known that the reaction path involves alternate oxidation and reduction of the active nickel centers, it has been established that in the usual operating range of heat flux [17,000 to 21,000 BTU/(hr)(sq ft)] the rapid reactions are controlled by the rate of heat transfer (1,3,5).

The chemical steps are very rapid causing low effectiveness factors, such that the reaction rates are proportional to the outside surface of the pellet (10). The Raschig-ring form of catalyst increases the utilizeable area and acts to reduce pressure drop.

In the aforementioned heat-flux range, the composition at the outlet of the converter corresponds to the equilibrium composition for reaction 3 at the outlet temperature and pressure and to the equilibrium composition of reaction 1 calculated at the outlet pressure and a temperature variously reported as $15–25°C$ below the actual (3,5,7). This equilibrium approach criterion is dependent on contact time and steam ratio. A conservative design value (3) is a ΔT of $50°F$ ($\sim 28°C$).

Design Calculations

It is possible in this case to ignore kinetics provided one confines economic studies within the narrow range of heat flux stated. Higher heat fluxes are now possible because of improved materials which permit high tube-wall temperatures. If these higher fluxes are used, the reaction could become more strongly rate limiting such that a simple equilibrium approach criterion would no longer be accurate or valid. In such a case a kinetic model would be required. For the heat flux range quoted above, however, a kinetic model would be a waste of time. The following algorithm based on equilibrium and heat transfer is adequate (3).

Algorithm

1. Set inlet composition, flow rates (including steam-to-methane ratio), and temperature. Set catalyst size, tube length, and tube diameter. Also select an average design heat flux. Set also the outlet temperature, pressure and approach to equilibrium.

2. Calculate the equilibrium constant K_1 for the stream reforming reaction at the outlet temperature minus ΔT approach.

$$K_1 = \exp(30.53 - 4.8486 \times 10^4/T + 2.421748 \times 10^6/T^2$$
$$+ 2.49 \times 10^9 T^3) \tag{CS-11.2}$$

3. Calculate the equilibrium constant K_3 at the reactor outlet temperature.

$$K_3 = \exp(-2.930632 + 3606.211/T + 5.0424 \times 10^6/T^2$$
$$- 1.815388 \times 10^9/T^3) \tag{CS-11.3}$$

4. Solve for corresponding compositions.

$$K_1 = \frac{(n_{H_2})_a{}^3(n_{CO})_a}{(n_{CH_4})_a(n_{H_2O})_a} \frac{P^2}{n_{T_a}{}^2} \tag{CS-11.4}$$

where $(n_{H_2})_a$ etc. are moles of indicated component per moles of equivalent methane fed and n_{T_a} is the total moles on same basis. [Since hydrocarbons other than methane are rapidly converted to methane near the inlet by hydrocracking, the initial or equivalent moles of methane should be (1), $\sum n \mathscr{F}_n$ for all values of n, where n is the number of carbon atoms in hydrocarbon, and $\mathscr{F}_n$ is the flow of hydrocarbon in moles t^{-1}.]

$$K_3 = \frac{(n_{H_2})_a(n_{CO_2})_a}{(n_{CO})_a(n_{H_2O})_a} \tag{CS-11.5}$$

Solve simultaneous Eqs. 11.4 and 11.5 by trial and error with $(n_{CH_4})_a = n$, $(n_{CO_2})_a = m$, $(n_{H_2})_a = \frac{1}{2}H - 3n - \overline{O} + 1 + 2m$, $(n_{H_2O})_a = \overline{O} - (1 - n) - m$, $(n_{CO})_a = 1 - n - m$, n_{T_a} = total moles including inerts, where $\overline{O}$ and H are equivalent atoms of oxygen and hydrogen in total feed per mole of equivalent methane fed.

5. Based on the selected tube size and length, and heat flux calculate number of tubes required.

$$\text{No. of tubes} = \frac{(\text{Heat Load})}{(\text{Heat Flux})\pi D Z_H}$$

where Z_H is the heated length (in this case 37 ft).

$$\text{Heat Load} = \sum_{j=1}^{n} \mathscr{F}_{j_e}(H_{f_j}{}^\circ + H_j - H_j{}^\circ)_e - \sum \mathscr{F}_{j_0}(H_{f_j}{}^\circ - H_j - H_j{}^\circ)_0$$

where $H_{f_j}^{\,\circ}$ is the heat of formation at $25°C$ for component j, H_j and $H_j^{\,\circ}$ are the enthalpies of component j at temperatures T and $25°C$, respectively. The outlet molar flow rates $(\mathscr{F}_{j_e})$ are determined from material balance based on equilibrium calculations.

6. Calculate G:

$$G = \frac{\text{mass flow/hr}}{(\pi D^2/4)(\text{number of tubes})}$$

7. Calculate pressure drop.

Equations for pressure drop on p. 491[1] may be used with a multiplying factor of 0.6 to account for the lower ΔP of Raschig-ring shapes (3). For this system the following special manufacturer's equation was used based on an average viscosity (11).

$$\Delta P = 5.922 \times 10^{-9} G^{1.9}\left(\frac{1-\varepsilon}{\varepsilon^3}\right)\frac{1}{\rho_f D_p^{1.1}} Z, \qquad \text{psi}$$

where D_p is the effective pellet diameter in inches, G is the mass velocity lb/hr ft^2, Z is the tube length in feet, and ρ_f is the fluid density in lb/ft^3.

Results

A major catalyst supplier provides a computer program based essentially on the previously described algorithm (11). It was used at typical operating conditions so it could be compared to published results as shown in Table CS-11.1. An equilibrium approach ΔT of $50°F$ was used and an average heat

Table CS-11.1 Comparison of Plant Operating and Calculated Data

	Plant Data (7)	Calculated
Tube length [heated length], ft	40[37]	[a]40[37]
Inside diameter, in	5	[a]5
Inlet pressure, atm	14.3	14.33
Outlet pressure, atm	12.2	[a]12.2
Gas inlet temp. °F	687	[a]687
Gas outlet temp. °F	1460	[a]1460
Feed carbon converted, %	91.7	90.7
No. of tubes	260	226
Approximate average heat flux, BTU/(hr)(sq ft)	17,000	17,000
Mass flow rate, lb/(h)(sq ft)	5476	6188

[a] Designates input data.

flux of 17,000 BTU/hr ft^2 was studied. The results are certainly within the errors associated with gathering plant data and in the model itself. The hydrocracking reactions for example, are exothermic but this heat is not considered in the model.

In applying the program, hydrocarbons other than methane are assumed to hydrocrack to methane at the initial portion of the tube. It has been observed that no hydrocarbon heavier than methane can be detected after contact with catalyst for a significant time (1). The fired length for heat transfer of a 40-ft tube was taken as 37 ft.

It appears that the method is quite adequate for design purposes. The 50°F approach is obviously conservative since a lower equilibrium approach would be required to reproduce the slightly higher observed conversion. The technique could be used to study the effect of pressure, inlet temperature, and steam-to-carbon ratio in order to determine the most economical allocation and recovery of energy, which is surely the major component in operating costs.

REFERENCES

1. M. H. Hyman, *Hydrocarbon Process.*, **47** (7), 131 (1968).
2. J. M. Moe and E. R. Gerhard, Paper No. 36d, 56th National Meeting American Institute Chemical Engineers, San Francisco, May 16–19, 1965.
3. *Girdler Catalysts*, Chemetron Chemicals Division of Chemetron Corporation, Louisville, Ky, 1965.
4. W. W. Akers and D. P. Camp, *A.I.Ch.E.J.*, **1** (4), 471 (1955).
5. G. W. Bridger, in *Catalyst Handbook*, Springer-Verlag, New York, 1970.
6. A. A. Khomenko, L. O. Apel'baum, F. S. Shub, Yu. S. Snagovskii, and M. I. Temkin, *Kinet. Katal.*, **12**, 423 (1971).
7. Yu. A. Sokolinskiy, M. Kh. Sosna, S. A. Markova, V. P. Semenov, and Yu. V. Shal'neva, *Khim. Prom.* (*Eng. Trans. The Soviet Chemical Industry*), **47**, 523, 478 (English Trans.) (1971).
8. O. A. Hougen, K. M. Watson, and R. Ragatz, *Chemical Process Principles*, Part 2, Wiley, New York, 1959.
9. F. A. Lee and K. D. Demarest, *Steam-Methane Reformer Furnaces*, Foster Wheeler Corp., June, 1967, New York.
10. D. W. Allen, E. R. Gerhard, and M. R. Likins, *Brit. Chem. Eng. & Proc. Tech.*, **17**, 605 (1972).
11. Chemical Products Division Chemetron Corp., now Girdler Chemicals, Inc., Louisville, Ky.

CASE STUDY 112

Vinyl Chloride Polymerization

THIS CASE is an example of the design of a large capacity batch reactor for multiple grades. Batch reactors are preferred because of the multiplicity of products together with varying market requirements. The large exothermic heat effects are controlled by using a suspension polymerization in water, which also aids in governing particle size along with adequate mixing and suspending agents. Selecting the optimum initiator and initiator concentration is also important in minimizing cycle time as well as spreading heat load over the cycle. The more complex problem of optimizing operating conditions has been discussed (12).

Major sources of uncertainty in this study are caused by lack of a precise values for heat of reaction and viscosity.

Problem Statement

Design a polyvinyl chloride polymerization reactor system for producing 100 million lb/yr of polyvinyl chloride

Monomer Specifications

Vinyl chloride	99.99+ %	Acetylenes,	
Suspended		acetalyde, etc.	30–50 ppm total
matter	None	Acidity	5 ppm HCl
Color	Water White	Water	100 ppm max.
Clarity	Clear	Methyl chloride	5 ppm
Allene	5 ppm max.	Heavies	
Iron	0.15 max.	(1,2 dichloro-	
1,3 butadiene	<5 ppm	ethane, etc.)	10 ppm max.

Products

Two major products, a flexible grade and a rigid grade are to be manufactured. Both will be required at the rate of 50 million lb/yr. Flexible PVC used in such products as films and squeeze bottles must accept plasticizer readily. This property is attained by limiting conversion to around 80%. When the pressure on the reactor is released at the end of the polymerization cycle, the unreacted monomer vaporizes from the polymer leaving a porous product into which plasticizer is readily incorporated. Rigid grade is, by contrast, made by carrying the conversion to 95% so that a relatively non-porous product remains which is ideal for rigid applications such as pipe. The character of these two major products is further varied by producing grades with different molecular weights as correlated with intrinsic viscosity.

Chemistry and Kinetics

Polymerization of polyvinyl chloride follows the classical free-radical mechanism with an initiator chosen of such activity and concentration that the reaction can be easily controlled at the desired operating temperature. Typical initiators are listed in Table CS-12.3 on p. 150.

The general free-radical mechanism applies.

$$d[I]/dt = k_d[I] \qquad \text{(CS-12.1A)}$$

Initiation $I \rightarrow 2R\cdot$ $(r_I) = 2f_e k_d[I] \qquad \text{(CS-12.1B)}$

where f_e is the initiator efficiency.

Propagation

$$R\cdot \quad + M \longrightarrow RM_j\cdot$$
$$RM_{n-1}\cdot + M \longrightarrow RM_n\cdot$$

$$r_{pr} = k_{pr}[M] \sum_{j=1}^{\infty} [RM] \qquad \text{(CS-12.2)}$$

Termination

$$RM_i\cdot + RM_j\cdot \longrightarrow RM_{i+j}R$$
$$RM_i\cdot + RM_j\cdot \longrightarrow RM_i + RM_j^=$$

$$r_t = 2k_t \left(\sum_{j=1}^{\infty} [RM_j\cdot] \right) \left(\sum_{i=1}^{\infty} [RM_i\cdot] \right) = 2k_t \left(\sum_{j=1}^{\infty} [RM_j\cdot] \right)^2,$$

or for convenience simply,

$$r_t = 2k_t[R\cdot]^2 \qquad \text{(CS-12.3)}$$

Transfer

$$RM_i\cdot + M \longrightarrow RM_i + M\cdot$$

The k_t term is defined in the model to follow (1) such that the factor 2 must be applied to express the rate of termination. This is a common mode for expressing k_t in Britain and Canada. Care must be exercised in using literature data, for in the U.S.A. k_t is usually given by $2k_t$.

The total concentration of all growing chains can be deduced from observables by invoking the Bodenstein steady-state approximation involving no change in free-radical concentration with time. Hence the rate of initiation must equal the rate of disappearance of radicals.

$$(r_1) = 2k_t[R\cdot]^2$$

$$[R\cdot] = \left[\frac{(r_1)}{2k_t}\right]^{\frac{1}{2}} = \left(\frac{f_e k_d[I]}{k_t}\right)^{\frac{1}{2}}$$

and

$$r_{pr} = \frac{-d[M]}{dt} = k_{pr}\left(\frac{f_e k_d}{k_t}\right)^{\frac{1}{2}}[I]^{\frac{1}{2}} \qquad \text{(CS-12.4)}$$

Thermodynamics

The reaction is one of the most highly exothermic polymerizations with ΔH variously reported as -22.6 to -27.5 kcal/g mole (6,10,11). The wide variation is attributable to the difficulty associated with calorimetry for polymerization reactions and variations in feedstocks and solid product. Obviously, precise design will require a more accurate value based on the particular product(s) of interest. For illustrative purposes the lower value has been used. This quantity should be defined more accurately for a definitive design. The high-temperature sensitivity and heat effects are emphasized by the high adiabatic factor and heat-generation potentials given in Table 6.4 (700 and 90.54, respectively).

Kinetics and Model Development

In the suspension polymerization of vinyl chloride, monomer is dispersed as droplets in the water. As polymerization occurs the droplets become gel-like in character, consisting of polymer-rich and monomer-rich portions.

The polymer is essentially insoluble in the monomer, but monomer is soluble in the polymer. Thus polymerization occurs in both phases. As long as a free monomer phase exists the equilibrium solubility of monomer in polymer remains constant (1). When the monomer phase is consumed, polymerization proceeds in the polymer-rich phase until the monomer dissolved in it disappears or the reaction is stopped. In the following model these distinct periods are defined analytically in terms of the conversion (X), as weight fraction converted being less or more than w_2 the weight fraction of polymer in the polymer-rich phase.

The drops behave in the same manner as in bulk polymerization of vinyl chloride, and the model is applicable to both bulk and suspension processes. In the suspension process water is of great value both in controlling temperature and ultimate polymer particle size. In addition to stirring speed various proprietary additives, some of which effect interfacial tension, are important in controlling polymer size and characteristics.

Model (1)

A model has been developed (1) that provides a means for estimating the time and temperature required for the desired molecular weight and conversion with a particular initiator and initiator concentration. Other factors of importance must be based on empirical and largely proprietary knowledge.

During the first period for which $X < w_2$, the monomer concentrations are constant in both phases (ρ_M/M_M in monomer) and $(1 - w_2)\rho_2/M_M$, where ρ_M and ρ_2 are the densities of monomer and polymer rich phase, respectively) and M_M is monomer molecular weight. The initiator is assumed to have the same concentration in both phases, and Eq. 12.1A is used to obtain an expression for its value at any time, t. The combined rates for both phases can then be written based on Eq. CS-12.4 with the rate constant for the polymer-rich phase $k_2 = pk_1$ where p is the empirical multiplier greater than unity which increases k_2 over k_1, the rate constant for the monomer-rich phase. This accounts for the increase in apparent rate constant caused by the decline in termination rate because of the restricted movement of growing polymer chains. The following equation is obtained (1).

$$\frac{dX}{dt} = k_1(w_1)_0^{\frac{1}{2}} \frac{1 + Q^\circ X}{(1 - BX)^{\frac{1}{2}}} \exp\left(-\frac{k_d t}{2}\right) \qquad \text{(CS-12.5)}$$

where $Q^\circ = [p(1 - w_2) - 1]/w_2$, $B = (\rho_{pr} - \rho_M)/\rho_{pr}$, $(w_1)_0$ is the initial weight fraction of initiator in monomer at $X = 0$, k_d is the initiator decomposition rate constant, and $k_1 = k_{pr}(f_e k_d/k_t)^{\frac{1}{2}}(\rho_M/M_1)$. M_1 is the molecular weight of the initiator, ρ_M and ρ_{pr} are the densities of the monomer and polymer, respectively. With initial conditions of $X = 0$ at $t = 0$ the solution is (1),

$$t = -\frac{2}{k_t} \ln (1 - \mathscr{H})$$ (CS-12.6)

where

$$\mathscr{H} = \frac{k_d}{2k_1(w_1)_0^{\frac{1}{2}}}$$

$$\times \left[\frac{2}{Q^\circ} (\sqrt{1 - BX} - 1) + \frac{\sqrt{Q^\circ + B}}{Q^\circ\sqrt{Q^\circ}} \ln \left\{ \frac{\dfrac{\sqrt{Q^\circ(1 - BX)} - \sqrt{Q^\circ + B}}{\sqrt{Q^\circ} - \sqrt{Q^\circ + B}}}{\dfrac{\sqrt{Q^\circ(1 - BX)} + \sqrt{Q^\circ + B}}{\sqrt{Q^\circ} + \sqrt{Q^\circ + B}}} \right\} \right]$$

Since when $X = 0$, the right-hand side of Eq. CS-12.5 reduces to $k_1(w_1)_0^{\frac{1}{2}}$, k_1 can be evaluated as the initial slope of an X versus $t(w_1)_0^{\frac{1}{2}}$ plot.

For the period where $X > w_2$ (usually $X \approx 70$–80%) where the monomer phase has disappeared, not only the termination rate constant k_t, but also k_{pr} and f_e decline as diffusion control ensues. This complex phenomenon can be handled by expressing the decline in the apparent k as a function of unreacted monomer, $1 - X$, as follows (1):

$$k = pk_1 \frac{1 - X}{1 - w_2}$$

This form agrees with experimental evidence that suggests a decline in the apparent rate constant with conversions of $X > w_2$.

The corresponding rate of monomer disappearance by a similar development as for Eq. CS-12.5 is given by

$$\frac{dX}{dt} = \frac{pk_1}{1 - w_2} (w_1)_0^{\frac{1}{2}} \frac{(1 - X)^2}{(1 - BX)^{\frac{1}{2}}} \exp \left(-\frac{k_d t}{2} \right)$$ (CS-12.7)

when $X > w_2$.

The parameter Q° is selected as the adjustable parameter for forcing the equation to fit experimental data, and p is determined from it. Values of Q° and w_2 determined from experimental data (1,2) are presented in Table CS-12.1 along with an equation for the rate constant k_1.

Integrating Eq. CS-12.7

$$t = \frac{-2}{k_d} \ln(1 - \mathscr{H}') + t_2$$ (CS-12.8)

Table CS-12.1   Values of Parameters from Experimental Data

Temp. °C	w_2	$Q°$	Data Source Ref.
30	0.80	4.4	1
50	0.77	5.0	1
65	—	6.1	2
70	0.72	5.4	1

Empirical equations based on these data are:

$$Q° = 2.4176 + 0.07769T_y - 0.0004371T_y^2$$
$$w_2 = 0.8075 + 0.0005T_y - 0.000025T_y^2$$

where $T_y =$ °C.
Rate constant (1):

$$(k_1)_A = 3.54 \; 10^{11}e^{-8505/T}, \text{hr}^{-1} \text{(mole fraction)}^{-1}$$

where $T =$ °K.

This is based on AIBN initiator, designated by sub A, (1) for which $k_d = 3.79 \times 10^{18} e^{-15,460/T}$, hr^{-1} (see Table CS-12.3).

Hence for any initiator

$$k_1 = k_d^{\frac{1}{2}}\left(\frac{k_1}{k_d^{\frac{1}{2}}}\right)_A$$

---

where t_2 is the time to reach $X = w_2$.

$$\mathcal{H}' = \frac{(1 - w_2)k_d}{2pk_1(w_I)_0^{\frac{1}{2}}}\frac{\sqrt{1 - BX}}{1 - X} - \frac{\sqrt{1 - Bw_2}}{1 - w_2}$$
$$+ \frac{B}{2\sqrt{1 - B}} \ln \frac{\left(\dfrac{\sqrt{1 - X} - \sqrt{1 - B}}{\sqrt{1 - w_2} - \sqrt{1 - B}}\right)}{\left(\dfrac{\sqrt{1 - X} + \sqrt{1 - B}}{\sqrt{1 - w_2} + \sqrt{1 - B}}\right)}$$

Experimentally it has been shown that the values of $\mathcal{H}$ and $\mathcal{H}'$ are small such that

$$\ln (1 - \mathcal{H}) = -\mathcal{H} - \tfrac{1}{2}\mathcal{H}^2 - \tfrac{1}{3}\mathcal{H}^3 \cdots$$
$$\approx -\mathcal{H}$$

Hence at $X < w_2$

$$t = \frac{2}{k_d}\frac{f(X)}{(w_I)_0^{\frac{1}{2}}} \tag{CS-12.9}$$

and at $X > w_2$

$$t = \frac{2}{k_d} \frac{f(w_2)}{(w_1)_0^{\frac{1}{2}}} + \frac{2g(X)}{k_d(w_1)_0^{\frac{1}{2}}} \qquad \text{(CS-12.10)}$$

$X = w_2$ can be found on a plot of X versus $t(w_1)_0^{\frac{1}{2}}$ as the point of maximum slope.

Molecular Weight

Experimental evidence (1–3) seems to confirm the dependency of molecular weight solely on temperature, as would be suggested by termination of living polymer occurring largely by chain transfer to monomer (4). Although disagreement exists, it seems reasonable, in view of the experimental findings and the nature of the growing polymer particles that occlusion would favor primarily chain transfer which, in turn, would be a function of temperature and not initiator concentration or conversion.

Accordingly, excellent data (1–3) for various types of initiators and polymerization recipes was fitted to a second-order equation in temperature.

$$\overline{M}_n = 158200 - 2427T_y + 7.599T_y^2 \qquad \text{(CS-12.11)}$$

where $T_y = {}^\circ C$.

The molecular weight distribution has been shown to follow the most probable distribution as would be expected by this terminating mechanism (3).

Molecular weight is correlated with the more measurable parameter of inherent viscosity η_I, (ASTM D1243–66). These data can be fitted to an expression second order in $\overline{M}_n$.

$$\eta_I = 19200 + 19.1259 \overline{M}_n - 1.74210^{-5}\overline{M}_n^2 \qquad \text{(CS-12.12)}$$

Equations CS-12.12 and CS-12.13 can be combined to produce a second-order equation in η_I relating η_I to $T_y(^\circ C)$.

$$T_y = 90.192 - 42.537 \eta_I + 3.512 \eta_I^2 \qquad \text{(CS-12.13)}$$

Design Conditions

Operating Temperature and Conversion

Equation CS-12.13 can be used to determine the operating temperature for the four grades planned. High conversions are required for the rigid grades (95%), but lower conversions are needed for the flexible grade (80%), as discussed on p. 140.

Summary of Operating Requirements

Grade	Inherent Viscosity	Required Temp. °C	Conversion X, %
Rigid	0.75	60.26	95.0
	1.0	51.16	95.0
Flexible	0.90	54.75	80.0
	1.00	51.16	80.0

Vessel and Impeller Selection

Impeller. An open, flat-blade turbine will be used to ensure high shearing action for dispersing monomer in water and keeping polymer particles dispersed. Four blades are selected for this design, but the economics of six blades should also be studied based on manufacturer's horsepower, pumping capacity, and cost for both.

Vessel Type. A conventional jacketed kettle will be used with electro-polished stainless-steel cladding to prevent polymer sticking. Water-cooled baffles will not be specified. The clad vessel has a higher thermal conductivity than solid stainless steel which has an effect on the overall U. The mixer shaft will enter from the bottom as shown in Fig. 8.1, p. 333[l]. This eliminates the unwetted portion of the shaft in a top-entering design that becomes rapidly covered with polymer because it is not washed continuously by the reaction mixture. Of course, the bottom entering mixer is much easier to support. Four reactor sizes will be considered as shown in Table CS-12.2.

Water-to-Monomer Ratio

The weight ratio of water-to-monomer used varies from 1.3 to 4 : 1 (7,9). By using less water more monomer can be polymerized per batch, but the heat load is, of course, higher. A careful choice of initiator, however, should permit operating at low water ratios so that production per unit of capital investment will be maximized. A value of 1.5 was selected for this study.

Overall Heat-Transfer Coefficient

As is true with most polymer systems, the most reliable data are based on actual operating data. Overall coefficients for carbon steel clad with electro-polished stainless steel are reported to be 75–130 BTU/hr ft^2 °F (9). The higher conductivity of the carbon steel portion and high jacket-water velocities assure high coefficients.

Table CS-12.2 Reactor Alternates: Heat Transfer and Horsepower

Rated Capacity, Gallons	Actual[a] Capacity, Gallons	Reactor Dimensions, In.			Calculated hp	Standard Motor Size, hp	h_i	Thickness Carbon Steel, In.	U
		D_1,[b] In.	ID	Tangent to Tangent					
5000	5136	45	108	108	41	50	151	0.5625	95
7500	8210	52	126	126	65.5	75	148	0.6694	92
10,000	10,715	57	136	136	85.4	100	146	0.7458	89
15,000	15,696	57	144	192	124.8	125	146	0.7917	88

[a] Operate at 90% of actual capacity so that essentially all of the heat transfer area is used.

[b] $\dfrac{D_{1_2}}{D_{1_1}} = \left(\dfrac{V_2}{V_1}\right)^{\frac{1}{3}}$ except for 15,000 gal. case. This largest size may require dual impellers.

Using Eq. 8.14, p. 357[1], an apparent viscosity of 50 cp, which has been reported (8), and a tip speed of 1100–1200 ft/min (Table 8.10) for which $N = 100$ rpm for 5000-gal tank, a value of h on the agitated side of 151 BTU/hr ft^2°F was estimated based on the following properties at 50°C. $w_2 = 0.7$, $\bar{\lambda} = 1.028 \times 10^{-3}$ estimated from $\lambda_f = 1.52 \times 10^{-3}$, $\lambda_{pr} = 0.39 \times 10^{-3}$ (5) and volume fraction suspended $= \bar{\rho}/(1 + m_f)\rho$, where $\bar{\rho}$ is the average density of reaction mass $= (1/\rho) + m_f/\rho_{H_2O}$, ρ is the average density of polymer-monomer mixture at $w_2 = 0.7$, $w_2/\rho_{pr} + (1 - w_2)/\rho_M$, $\rho_{pr} = 1.4$ g/cm^3 (6), $\rho_M = 0.8$ (6), $\bar{c}_p = 0.7$ (5), and m_f is the mass ratio of water to monomer in initial charge. For $\frac{9}{16}$ in. carbon steel and $\frac{1}{8}$ in. stainless steel

$$\frac{1}{U} = \frac{1}{h_o} + \frac{1}{h_i} + \frac{0.125}{105} + \frac{0.5625}{360}$$

where conductivity of stainless $= 105$ and carbon steel $= 360$ BTU/hr (ft^2/in.)°F, h_o is the jacket-side coefficient and h_i is the agitated side coefficient. From Table 8.7 for jacket

$$D_H = 4J_w = (4)(1.5) = 6 \text{ in}$$

h_o @ 8 ft/sec water velocity and 50°F from Appendix F.

$$h_o = (1400)\left(\frac{0.62}{6}\right)^{0.2} = 899$$

$$\frac{1}{U} = 0.00388 + \frac{1}{h_i}$$

For $h_i = 151$, $U = 95$ BTU/hr ft^2°F.

Values for the three other reactor sizes can be estimated by a scale-up procedure. Thickness of the carbon steel increases linearly wtih diameter, and the value of h_i was scaled on the basis of equal power per unit volume which from Table 8.11 requires

$$\frac{(h_i)_2}{(h_i)_1} = \left[\frac{(D_1)_2}{(D_1)_1}\right]^{1.33m - 1} \qquad m = 0.65$$

These estimated heat-transfer coefficients shown in Table CS-12.2 are consistent with the range of reported values and will be used for illustrative purposes. Since a four-blade impeller is finally selected, some adjustment in the heat-transfer coefficient just calculated is indicated but not possible with any accuracy using literature data. Manufacturer's recommendations on the particular agitator selected should be sought.

Power Requirements

Power estimates were based on the open four-blade turbine. Maintaining the monomer droplets in the proper state of dispersion, although aided greatly by various proprietary additives, requires reasonably high shear rate as well as adequate turn over of the vessel contents. Cleaning is easier for an impeller with open construction. For the 5000-gal reactor

$$(N_{Re})_l = \frac{\bar{\rho}ND_I^2}{\mu} = \frac{(1.06)(100/60)(45 \times 2.54)^2}{(50)(0.01)} = 46161$$

From Fig. 8.7 (curve 2), $N_p = (4)(4/6)^{0.8} = 2.89$

$$P_a = \frac{(2.89)\bar{\rho}N^3D^5}{g_c} = \frac{(2.89)(1.06)(62.4)\left(\dfrac{100}{60}\right)^3\left(\dfrac{45}{12}\right)^5}{32.17}$$

$$= 20400 \text{ ft lb}_f/\text{sec or } \frac{204000}{550} = 37 \text{ hp}$$

This is $\frac{37}{5} = 7.4$ hp/1000 gal which is close to that recommended (6–7 hp/1000 gal) for such units (10).

For other sizes considered the scale-up will be on the basis of equal power per unit volume, realizing that some adjustment in additives may be necessary to assure comparable product particle size. The total hp required for the 5000-gal reactor is 41.2 when a 10% driver loss and a 0.5 hp seal loss are added. A standard 50-hp explosion-proof motor is selected. The results for all sizes are summarized in Table CS-12.2.

Selection of Initiator and Cooling Temperature

The production rate of a batch polymerization requiring careful temperature control is limited by the heat removal potential of the jacketed reactor. Thus an initiator that will approach a uniform heat-generation rate is preferred over one that results in a single large peak.

Vinyl-chloride polymerization, because of the decline in the termination rate constant in the first period, increases in rate with conversion and time up to a maximum at the point $X = w_2$. The form of this increase is governed by the value of the initiator rate constant at the temperature of polymerization. If the value is low, the reaction will begin slowly and more initiator will be required to complete the reaction in a reasonable time period. Higher values will produce faster initial rates, require low concentrations, and produce more even heat generation. This can be seen analytically by reference to Eq. CS-12.5 in which k_d has an accelerating effect initially but moderates the reaction at longer times, $k_1 = \alpha k_d^{\frac{1}{2}}$.

These characteristics of various initiators are best compared by direct experimental observation or by using experimentally determined values of k_d in an equation such as Eq. CS-12.5. Table CS-12.3 presents a group of

Table CS-12.3 Comparison of Initiators at 60.26°C Reaction Temperature

Initiator	k_d @ 60.26°C hr^{-1}	$A(10^{-18})$ hr	E/R, °K [a]	M_I	Efficient Temp. Range, °C	Average Jacket Temp. 100°F (38°C)		Average Jacket Temp. 45°F (7.6°C)	
						Optimum % Initiator	Time, Hr @ 95% Conversion	Optimum % Initiator	Time, Hr @ 95% Conversion
AIBN	0.270	3.79	15460	164		0.0138	22.24	0.062	10.5
tert-butyl peroxypivalate	0.127	1.33	14595	174.2	50–65	0.0085	16.21	0.024	8.49
lauroyl peroxide	0.0555	2.628	15098	398.6	50–65	0.0231	19.9	0.087	9.50
di-isopropyl peroxydicarbonate	0.472	0.241	13588	206.2	45–55	0.0150[b]	>9.12	0.025	5.83
di(sec-butyl) peroxydicarbonate	0.493	0.251	13588	234	45–55	0.0175[b]	>8.90	0.028	5.73

[a] Evaluation of Organic Peroxides from Half-Life Data, Bulletin 30–30, Lucidol Division Pennwalt Corporation, Buffalo, New York, 1972.

[b] Not a true optimum. Maximum dq_g/dt occurs early.

Pressure = 130 psig with tank volume of 10715.

initiators commonly used in vinyl-chloride polymerization. Manufacturers will recommend the best initiator for the target operating temperature, for example, acetyl cyclohexyl sulfonyl peroxide is usually recommended for use below 50°C because at higher temperatures it is difficult to remove the heat generated with conventional systems. It will be instructive, however, to consider all five of these initiators by comparing the time required for the rigid grade in a 10,000-gal tank. This can be done by determining the time required for producing the desired conversion at the maximum initial initiator concentration which will not exceed the cooling capacity of the system. Since the time t_2 at which $X = w_2$ is affected by the initiator concentration, this calculation involves solving the following (see algorithm p. 154) which equates heat generation to heat removal.

$$\frac{m_{\mathrm{d}}(-\Delta H)}{M_{\mathrm{M}}}\frac{dX}{dt_2} = \frac{m_{\mathrm{d}}(-\Delta H)}{M_{\mathrm{M}}} k_1(w_1)_0^{\frac{1}{2}} \frac{1 + Q^{\circ}X}{(1 - BX)^{\frac{1}{2}}} \exp\left(\frac{-k_{\mathrm{d}}t_2}{2}\right)$$

$$= UA_{\mathrm{h}}(T - \bar{T_{\mathrm{j}}}) \tag{CS-12.14}$$

where m_{d} is the total mass of monomer in drops initially.

The temperature of the cooling medium must be selected first. Two possibilities exist for water, cooling tower water ($\approx 100°F$) and chilled water ($\approx 45°F$). For these purposes calculations are made with several initiators using the algorithm on p. 154. Because of the reduced ΔT for cooling-tower water, the heat removal rate is reduced and lower initiator concentration must be used. The result is excessively long reaction times (see Table CS-12.3) even for the most efficient initiators.

The maximum heat generation in these preliminary calculations for the chilled-water case is approximately 4.7 MM BTU/hr. With the 5000-gal reactor, a 2.3-in. space exists for the jacket. Using a 16.7-in. baffle spacing and 8 ft/sec water velocity, the minimum temperature rise of the coolant can be calculated. Higher rises and lower flow rates will be used during the early portion of the cycle when the heat removal duty is less.

$$(\Delta T)\frac{(8)(3600)(2.3)(16.4)(62.7)}{144} = 4.7 \times 10^6$$

$$\Delta T = 10°F$$

Thus a reasonable average $\bar{T_{\mathrm{j}}} = 40 + \Delta T/2 = 45°F$ or 7.6°C.

With this value of $\bar{T_{\mathrm{j}}}$ the five initiators are compared at 60.26°C operating temperature and 95% conversion in Table CS-12.3 and Figs. CS-12.1 and 2. The two peroxydicarbonates appear ideal for this system. The heat release is more evenly distributed and the reaction time shorter than for lauryl peroxide or *tert*-butyl peroxypilvalate. Lauryl peroxide, in particular, starts slowly and the major heat load occurs over a shorter period of time. Thus the di (*sec*-butyl) peroxydicarbonate will be selected.

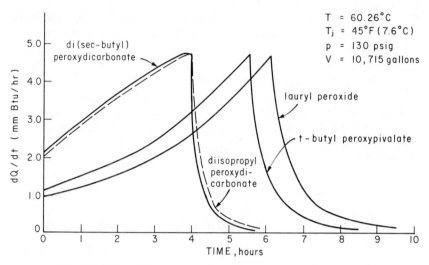

Fig. CS-12.1   Initiator comparison: Heat generation versus time.

Since the 45°F average chilled-water temperature produces the maximum heat removal, the cooling temperature will have to be allowed to rise to higher values at points other than the maximum so that the reaction will not be slowed. This can be accomplished by lowering the chilled water flow rate. Initially, the reaction mass must be heated to the reaction temperature. This is best done in a premix tank in which all the ingredients but the initiator are added, mixed, and preheated. In this way batch time is not lost in preheating or mixing.

Operating Pressure

The operating pressure will be set at the combined vapor pressure of vinyl chloride and water corresponding to each operating temperature, Deviations from ideality are not significant. These pressures are 105.2 psig at 51.2°C, 115.5 psig at 54.8°C, and 133.2 psig at 60.3°C.

Design Study

Using di(sec-butyl) peroxydicarbonate as initiator and temperatures and pressures necessary for the several grades, the four reactor sizes given in Table CS-12.2 can be studied. The time required for each grade and number of reactions of a particular size can be determined from the equations already presented assuming two trains of reactors, one for flexible grades and one for rigid grades.

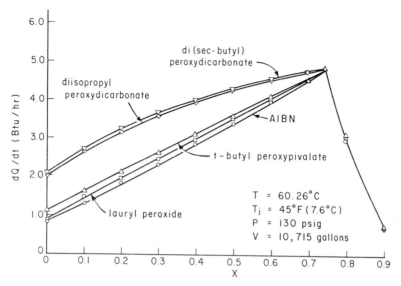

Fig. CS-12.2 Initiator comparison: Heat generation versus conversion.

The following times for cleaning and removing product apply (see also Ref. 9)

Reactors are cleaned every 4 batches

Cleaning time: 6 hr

Time for stripping monomer and recharging after reaction: 30 min*
Thus time for 4 batches $= 4t + 6 + (4)(\frac{1}{2}) = 4t + 8$

$$m_{pr} = \frac{4m_d X}{4t + 8} = \frac{m_d X}{t + 2} \qquad \text{(CS-12.15)}$$

where m_d is the mass of monomer in total droplets initially, and m_{pr} is the mass of polymer produced per unit time.

m_d is related to the total monomer charge m_M: $m_M = m_d + w_M m_w + \rho_v(V - V_w - V_d)$ where m_w is the mass of water, w_M is the mass monomer

---

* This short time can be realized by providing a premix tank (one for each train) in which monomer, water, and additives but no initiator are mixed and preheated (9). A circulating pump can be used together with an external heat exchanger. The mix can be added to the reactor at the proper temperature, the agitator turned on and the initiator added with minimal use of the 30 min alloted. The major portion of this time is consumed after the reaction in releasing the pressure to around 2 psig, followed by evacuating with a vacuum pump to 15 in. of Hg. Steam can be added toward the end of the cycle to further strip the polymer. Discharge of polymer slurry from the reactor is by gravity.

dissolved/mass of $H_2O = 0.015$ (2), ρ_v is the density of vapor, and V, V_w, and V_d are the reactor, water, and drop volumes, respectively. Assuming the vapor is primarily monomer, $\rho_v = PM_M/RT$, and $V_d = PM_M/\rho_M RT$

$$m_d = \frac{m_M - w_M m_w - \dfrac{PM_M}{RT}\left(V - \dfrac{m_w}{\rho_w}\right)}{1 - \dfrac{PM_M}{\rho_M RT}} \qquad \text{(CS-12.16)}$$

Using a ratio of water to monomer of 1.5 and assuming 90% of reactor full

$$0.9V = \frac{m_M}{\rho_M} + \frac{1.5 m_M}{\rho_w}$$

or

$$m_M = \frac{0.9V}{\dfrac{1}{\rho_M} + \dfrac{1.5}{\rho_w}} \qquad \text{(CS-12.17)}$$

Algorithm

The temperature dependent parameters were calculated and then the design parameters m_M (Eq. CS-12.17) and $m_w = 1.5\, m_M$. The value of m_d was then determined from Eq. CS-12.16. The initiator concentration was optimized by starting with $t_2 = 5$ (call t_2') and solving for $(w_I)_0$ from Eq. CS-12.14, followed by calculating t_2 from Eq. CS-12.6 using that value of $(w_I)_0$. If $t_2 - t_2' \nless 0.001$, t_2 is used in Eq. CS-12.14 and the cycle repeated. With the required initiator concentration thus determined t is calculated using Eqs. CS-12.6 and CS-12.8. Finally, heat generation rates for various values of t were determined,

$$\frac{dq_g}{dt} = \frac{(-\Delta H)(m_d)}{M_M}\frac{dX}{dt}.$$

Results and Final Design

The results of these studies for four sizes of reactors are summarized in Table CS-12.4. Two trains will be used so that flexible and rigid grades can be isolated. Equation CS-12.15 is used to calculate hourly production of each grade and type for each reactor size.

Based on an operating factor of 0.9, which includes time for changing types within a grade, the total operating time including cleaning, stripping, and recharging is

$$t_I + t_{II} = t_{III} + t_{IV} = (0.9)(365)(24)$$
$$= 7900 \text{ hr} \qquad \text{(CS-12.18)}$$

where $t_I, t_{II}, \cdots$ are the times required to produce a given product. Using these times and the annual production requirements the following balances apply

$$n_F t_I m_{pr_1} = 12.5 \times 10^6 \qquad \text{(CS-12.19)}$$

$$n_F t_{II} m_{pr_2} = 37.5 \times 10^6 \qquad \text{(CS-12.20)}$$

$$n_R t_{III} m_{pr_3} = 37.5 \times 10^6 \qquad \text{(CS-12.21)}$$

$$n_R t_{IV} m_{pr_4} = 12.5 \times 10^6 \qquad \text{(CS-12.22)}$$

where n_F and n_R are the numbers of reactors in flexible and rigid-grade trains, respectively. Equations CS-12.19–CS-12.22 may be solved for the respective times that can be substituted into Eq. CS-12.18 to yield

$$n_F = \left(\frac{1.58}{m_{pr_1}} + \frac{4.75}{m_{pr_2}}\right)10^3 \qquad \text{(CS-12.23)}$$

$$n_R = \left(\frac{4.75}{m_{pr_3}} + \frac{1.58}{m_{pr_4}}\right)10^3 \qquad \text{(CS-12.24)}$$

The results of these calculations are also included in Table CS-12.4.

Power Consumption

$$(\text{BHP-hr})_{I, II} = \text{BHP } n_F t_I\left(\frac{t}{t+2}\right)_I + \text{BHP } n_F t_{II}\left(\frac{t}{t+2}\right)_{II}$$

$$= \text{BHP}\left[\frac{12.5 \times 10^6}{m_{pr_1}}\left(\frac{t}{t+2}\right)_I + \frac{37.5 \times 10^6}{m_{pr_2}}\left(\frac{t}{t+2}\right)_{II}\right]$$

$$(\text{BHP-hr})_{III, IV} = \text{BHP } n_R t_{III}\left(\frac{t}{t+2}\right)_{III} + \text{BHP } n_R t_{IV}\left(\frac{t}{t+2}\right)_{IV}$$

$$= \text{BHP}\left[\frac{37.5 \times 10^6}{m_{pr_3}}\left(\frac{t}{t+2}\right)_{III} + \frac{12.5 \times 10^6}{m_{pr_4}}\left(\frac{t}{t+2}\right)_{IV}\right]$$

where BHP is the calculated required brake horsepower for one reactor from Table CS-12.2.

Power cost/3 yr = (Total BHP-hr)(0.7457)($0.0077/KW)(3)

The final costs of vessels and appurtenances plus power based on a 3-yr payout are shown in Table CS-12.4. The 7500-gal case is selected over the

Table CS-12.4 Summary of Calculations and 1972 Costs

	Flexible Grade		Rigid Grade		Flexible Grade		Rigid Grade	
Nonimal Volume, Gal.	5000				7500			
Actual Volume, Gal.	5136				8210			
U, BTU/hr ft^2°F	95				92			
	Type 1	Type 2	Type 3	Type 4	Type 1	Type 2	Type 3	Type 4
Reactor temp., °C	60.26	51.16	51.16	54.75	60.26	51.16	51.16	54.75
Conversion, %	80	80	95	95	80	80	95	95
Pressure, psig	133	105		115.5	133	105		115.5
Mass water, lb	21580	21774	same as Type 2	21698	34496	34806		34684
Mass monomer, lb	14387	14516		14465	22997	23204		23123
Mass monomer in drops, lb, (m_d)	13953	14098	same as Type 2	14041	22304	22536	same as Type 2	22445
Optimum percent initiator (based on monomer)	0.03658	0.04084		0.03775	0.03247	0.032		0.03012
Monomer phase disappears at $X =$	0.7469	0.7677		0.7599	0.7469	0.7677		0.7599
Maximum rate of heat generation MM BTU/hr	3.089	2.569		2.812	4.302	3.405		3.681

156

Time required, hr. (one batch)	3.335	5.125	7.14	6.104	3.658	5.965	8.269	7.055
Hourly production m_{pr}, lb/hr-reactor	2092	1583	1465	1646	3154	2264	2085	2355
No. of reactors and (actual number selected)	3.756(4)			4.20(5)	2.60(3)			2.95(3)
Unit cost $			69,000				87,650	
Total reactor cost, $			621,000				525,890	
BHP/yr	851,776			1,054,173	980,530			1,128,909
Total BHP-hr/yr			1,905,949				2,109,439	
Power cost/3 yr, $			32,831				36,336	
Total cost/3 yr, $			653,831				562,226	

Table CS-12.4 (*Continued*)

	10000 10715 89				15000 15696 88			
Nominal Volume, Gal. Actual Volume, Gal. U, BTU/hr ft^2°F	Flexible Grade		Rigid Grade		Flexible Grade		Rigid Grade	
	Type 1	Type 2	Type 3	Type 4	Type 1	Type 2	Type 3	Type 4
Reactor temp., °C	60.26	51.16	51.16	54.75	60.26	51.16	51.16	54.75
Conversion, %	80	80	95	95	80	80	95	95
Pressure, psig	133	105		115.5	133	105		115.5
Mass water, lb	45021	45425		45267	65950	66542		66310
Mass monomer, lb	30014	30284		30178	43967	44361		44207
Mass monomer in drops, lb (m_d)	29109	29412	same as Type 1	29294	42641	43085	same as Type 1	42912
Optimum percent Initiator (based on monomer)	0.02845	0.02755		0.02652	0.0258	0.02346		0.02309
Monomer phase disappears at $X =$	0.7469	0.7677		0.7599	0.7469	0.7677		0.7599
Maximum rate of heat generation MM BTU/hr	4.756	3.937		4.255	6.077	5.026		5.417

158

Time required, hr. (one batch)	4.074	6.564	9.067	7.676	4.432	7.296	10.036	8.440
Hourly production m_{pr}, lb/hr-reactor	3834	2748	2525	2876	5304	3708	3401	3905
No. of reactors and (actual number selected)		2.14(3)		2.43(3)		1.58(2)	1.80(2)	
Unit cost $		104,000					123,700	
Total reactor cost, $		624,000					494,835	
BHP/yr		1,080,097		1,333,661		1,193,367	1,470,469	
Total BHP-hr/yr,		2,413,758					2,663,836	
Power cost/3 yr, $		41,579					45,886	
Total cost/3 yr, $		665,759					540,721	

15,000-gal case because of the added flexibility of three reactors in each train over two and the relatively small difference in 3-yr costs. This decision needs further study. Power costs are higher for the 15,000-gal case and are increasing rapidly with escalating energy costs. Cleaning labor costs and maintenance that have not been included should be lower for the larger reactors. A three-blade curved-blade agitator with lower power requirements designed for large tanks has been reported (1). This change could make the 15,000-gal alternate more attractive, but it will require tests for optimum amounts and kinds of suspending agent to assure acceptable particle sizes. The incentive is great, however, to reduce power costs.

The reactors will be elevated so that the product slurry can be discharged by gravity to the dewatering and drying section. Each reactor will be provided with a rupture disc for possible runaways. Means for rapidly lowering pressure by discharging into the monomer recovery system will also be provided. The ensuing rapid evaporation of monomer will cool the mixture fast enough to moderate the runaway in most cases and prevent bursting of the rupture disc.

REFERENCES

1. A. H. Abdel-Alim and A. E. Hamielec, *J. Appl. Polym. Sci.*, **16**, 783 (1972).
2. A. F. Hauss, *J. Polym. Sci.*, Part C, **33**, 1 (1971).
3. J. Lyngaae-Jorgensen, *J. Polym. Sci. Part C*, **33**, 39 (1971).
4. J. Cotman, M. Gonzalez, and H. Claver, *J. Polym. Sci.*, *A-1*, **5**, 1137 (1967).
5. S. A. Miller, *Acetylene*, Vol. II, Academic, New York, 1966.
6. L. G. Shelton, D. E. Hamilton, and R. H. Fisackerly, in *Vinyl and Diene Monomers*, Vol. XXIV, Part 3, *High Polymers*, E. C. Leonard (ed.), Wiley-Interscience, New York, 1971.
7. H. A. Sarvetnick, *Polyvinylchloride*, Reinhold, New York, 1969.
8. *Seminar on Design Safety in Polymerization*, Pfaudler Co. Rochester, New York, April 26–27, 1966.
9. L. F. Albright, *Processes for Major Addition-Type Plastics and Their Monomers*, McGraw-Hill, New York, 1974.
10. W. F. Schlegel, *Chem. Eng. (N.Y.)*, p. 88 (March 20, 1972).
11. G. Beckmann, *Chem. Tech.*, p. 304 (May, 1973).
12. A. Pierru and C. Alexandre, *Hydrocarbon Process.*, **52** (6), 97 (1973).

CASE STUDY 113

Batch Hydrogenation of Cottonseed Oil

THIS CASE STUDY illustrates the modeling of a semibatch reactor for a gas–liquid–solid catalyst system involving a complex reaction scheme. Although the interplay of mass transfer and reaction are well illustrated by the model, further development including quantifying temperature effects and comparing with operating data are necessary. The model is used in this design to aid in estimating cycle time and maximum heat load, both of which would be more accurately known from existing plant data.

A model of this type, when further developed, could be useful in guiding operating studies and as an aid in process control of this semibatch process.

Problem Statement

Size and select the number of batch stirred tanks required for hydrogenating cottonseed oil to various products required for margarine and shortening. A production rate of 9×10^5 to 1×10^6 lb/day is required. Rufert nickel catalyst composed of 24–26% nickel dispersed in hard stearin is to be used. The feedstock will be highly refined, and the catalyst will be used once and discarded.

Feed Composition as Acid

	Wt. %	No. Carbon Atoms	Double Bonds
Saturated acids (S)	25.6	16–18	0
Cis-oleic acid (R_1)	27.0	18	1
Iso-oleic acid (trans)(R_2)	0.4		
Linoleic acid (B)	47.0	18	2
	100.0		

$$C_{B_0} = 1.45 \text{ g moles/l.}$$
$$C_{R_1} = 0.83 \text{ g moles/l.}$$

General Structure

$$
\begin{array}{l}
\quad\quad\quad O \\
\quad\quad\quad \| \\
CH_2-O-C-R \\
\quad\quad\quad O \\
\quad\quad\quad \| \\
CH-O-C-R' \\
\quad\quad\quad O \\
\quad\quad\quad \| \\
CH_2-O-C-R''
\end{array}
$$

The typical fatty-oil trigylceride will yield glycerine on hydrolysis, and the several fatty acids represented by the R groups. Commercial fats are, therefore, described in terms of fatty-acid content. Simple hydrogenation as conducted for producing margarine and shortening merely saturates double bonds and isomerizes. It does not hydrolyze the ester. Results of the hydrogenation are described in terms of the degree of saturation expressed as an iodine value (number of grams of iodine that react with 100 g of triglyceride oil) and the composition of the side chains as fatty acids. Typical iodine values for margarine and shortening are in the range of 65–80 (1).

Process Description

Oils such as cottenseed, soy, and corn oil are hydrogenated to reduce color and odor, improve stability, and to increase melting point so as to produce a solid or semisolid product at room temperature (1). These several goals are attainable by reducing the unsaturation, but nutritional studies indicate that polyunsaturated products are helpful in minimizing blood cholesterol levels. *Trans*-monounsaturated isomers have higher melting points. Conditions that favor high selectivity to monosaturates also tend to favor *trans*-isomers that are important in producing the desired consistency and good low-temperature characteristics.

Good products can also be produced by mixing a higher melting more saturated product with a liquid oil containing high amounts of acid groups with two and three pairs of double bonds such as safflower, soy, and corn oils.

A typical batch dead-end process is shown in Fig. CS-13.1, and steps in the process have been described (1–4). A portion of the oil charge which is weighed into the supply tank is used to mix with the catalyst in the mix tank. The oil is loaded to the reactor under vacuum (catalyst = 0.1 wt % per batch) and the agitator and heating are started. When a temperature about 50°C below the reaction temperature is reached, catalyst is loaded,

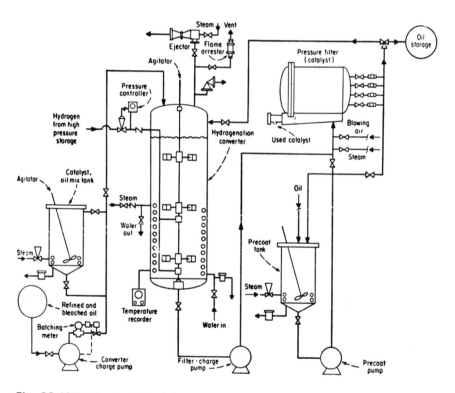

Fig. CS-13.1 Dead-end batch hydrogenation process for triglyceride oils. Reproduced by permission: L. F. Albright, *Chem. Eng.* (*N.Y.*), p. 249 (Oct. 9, 1967).

vacuum turned off, and hydrogen is added. Cooling water is turned on, and the heat of reaction increases the system temperature to the desired reaction temperature (see Chapter 12 for a description of such unsteady-state periods). When the reaction pressure has been reached, only enough hydrogen is added to maintain pressure. When the reaction has been completed, hydrogen flow is discontinued and the product removed, filtered and bleached.

Operating conditions vary depending on feed and desired product. Normal ranges are 130–200°C and 5–60 psig (2). The reactor operates initially in the semibatch mode as hydrogen is added at relatively high rates until the pressure builds up to the desired value. After this time, only a small amount of hydrogen is added to maintain the pressure, a decline of which is caused by hydrogen consumption and a bleed gas stream operated to expel impurities. In the second mode hydrogen continues to bubble but at a low rate. The impeller near the surface (see Fig. CS-13.1) will also draw some hydrogen from the vapor space and cause further contacting.

The end point can be detected approximately by refractive index, which correlates with iodine value or by noting hydrogen consumption (2). If the end point is critical, the agitator can be stopped to allow time for more thorough laboratory work such as iodine value and isomer analysis. This procedure is time consuming. Alternatively, batches of differing end points can be blended to produce the desired product.

Chemistry

The partial hydrogenation of natural vegetable oils is a most involved reaction involving both complex chemistry and mass transfer limitations. The most generally accepted simple mechanistic scheme is as follows (1,5,6):

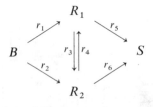

where B is the diunsaturated group, R_1 is the *cis*-monosaturated group, R_2 is the *trans*-monosaturated group, and S is the saturated group.

The actual mechanism is much more complex. For example, the isomerization shown is postulated to occur in a series of steps (8). Low hydrogen concentration encourages isomerization rather than complex hydrogenation. High reaction temperature, low pressure, high catalyst concentration, and low agitation rate decrease hydrogen concentration and thus increase the rate of isomerization over the rate of hydrogenation. These same conditions improve selectively defined as the ratio of the rates of hydrogenation of polyunsaturates to that for monounsaturates. The complex isomerization and saturation mechanistic scheme (5,8) involves a single activated hydrogen atom which attacks the double bond to yield an unstable partially unsaturated complex. If the concentration of hydrogen is high on the catalyst surface, the complex reacts with another hydrogen atom to saturate the bond. If hydrogen concentration is low, the complex will decompose to reform the double bond in the same or another position. Either the *cis* or *trans* form is produced so that this mechanism explains the observed positional and geometrical isomerization. The reaction scheme suggested is (5,8)

$$H_2 + 2\ell \underset{k_H'}{\overset{k_H}{\rightleftharpoons}} 2H\ell$$

$$B + \ell \underset{k_B'}{\overset{k_B}{\rightleftharpoons}} B\ell$$

$$B\ell + H\ell \underset{k_0'}{\overset{k_0}{\rightleftharpoons}} BH\ell + \ell$$

$$BH\ell + H\ell \xrightarrow{\bar{k}_1} R_1\ell + \ell \underset{k_{R_1}}{\overset{k_{R_1}'}{\rightleftharpoons}} R_1 + 2\ell$$

$$BH\ell + H\ell \xrightarrow{\bar{k}_2} R_2\ell + \ell \underset{k_{R_2}}{\overset{k_{R_2}'}{\rightleftharpoons}} R_2 + 2\ell$$

$$R_1\ell + H\ell \underset{\bar{k}_3'}{\overset{\bar{k}_3}{\rightleftharpoons}} RH\ell + \ell$$

$$R_2\ell + H\ell \underset{\bar{k}_4'}{\overset{\bar{k}_4}{\rightleftharpoons}} RH\ell + \ell$$

$$RH\ell + H\ell \xrightarrow{\bar{k}_5} S\ell + \ell \underset{k_S}{\overset{k_S'}{\rightleftharpoons}} S + 2\ell$$

where H, B, R_1, R_2, and S refer respectively to hydrogen and diunsaturated, *cis*-monounsaturated, *trans*-monounsaturated and saturated acid groups, ℓ is an active site, $RH\ell$ is complex formed from monounsaturated groups containing both *cis* and *trans* configurations.

Based on this scheme it now becomes clear why both selectivity (relative reactivity of polyunsaturates compared to monounsaturates) and isomerization of monosaturates are favored by low hydrogen concentration at the catalyst surface which is in turn encouraged by low agitator rpm, high temperature and low pressure. Diunsaturates are more strongly adsorbed than monounsaturates and are thus favored. If the hydrogen concentration is low on the catalyst, more diunsaturates can be adsorbed and react (1).

Further low hydrogen concentration encourages isomerization of mono-saturates rather than total saturation.

Kinetics

Rate equations have been derived based on the previously mentioned scheme and the following assumptions (5):

1. Adsorption and desorption steps at equilibrium.

2. The catalyst surface is sparsely covered by adsorbed components and thus the concentration of unoccupied sites (6) is essentially independent of the concentration in the liquid phase.

3. These equations apply to the time elapsed after the induction period. The induction period is defined by extrapolating the straight line portion of a plot of log iodine value versus time and for the conditions of this study is approximately 25 min (9).

$$\frac{dC_B}{dt} = -(k_1 + k_2)\sqrt{C_{H_s}}\,C_B \tag{CS-13.1}$$

$$\frac{dC_{R_1}}{dt} = k_1 C_{H_s}\sqrt{C_B} - k_3\sqrt{C_{H_s}}\,C_{R_1} + k_4\sqrt{C_{H_s}}\,C_{R_2} - k_5 C_{H_s} C_{R_1} \tag{CS-13.2}$$

$$\frac{dC_{R_2}}{dt} = k_2\sqrt{C_{H_s}}\,C_B + k_3\sqrt{C_{H_s}}\,C_{R_1} - k_4\sqrt{C_{H_s}}\,C_{R_2} - k_6 C_{H_s} C_{R_2} \tag{CS-13.3}$$

$$\frac{dC_s}{dt} = k_5 C_{H_s} C_{R_1} + k_6 C_{H_s} C_{R_2} \tag{CS-13.4}$$

where C_{H_s} is the concentration of hydrogen at catalyst surface, t is the pseudo-time, which is the actual elapsed time minus the initiation time,

$$k_1 = \bar{k}_1 k_0 C_l^2 K_B \sqrt{K_H}/(\bar{k}_1 + \bar{k}_2),$$

$$k_2 = \bar{k}_2 k_0 C_l^2 K_B \sqrt{K_H}/(\bar{k}_1 + \bar{k}_2),$$

$$k_3 = \bar{k}_3 k_4' C_l^2 K_{R_1} \sqrt{K_H}/(\bar{k}_3' + \bar{k}_4'),$$

$$k_4 = \bar{k}_4 k_3' C_l^2 K_{R_2} \sqrt{K_H}/(\bar{k}_3' + \bar{k}_4'),$$

$$k_5 = \bar{k}_5 k_3 C_l^2 K_{R_1} K_H/(\bar{k}_3' + \bar{k}_4'),$$

$$k_6 = \bar{k}_5 k_4 C_l^2 K_{R_2} K_H/(\bar{k}_3' + \bar{k}_4').$$

Analysis of extensive work on cottonseed oil at high agitator rpm (9) has established intrinsic values of the kinetic constants at 130°C and 0.07%

nickel charged. The pressure range is 20–140 psig. The values are $k_1 = 0.254 \, l.^{0.5}/(\text{g mole})^{0.5}(\text{min})$, $k_3/k_4 = 3.0$, $k_5 = k_6$.

Model

A realistic model for the actual case where mass-transfer resistance to hydrogen transfer is significant has been developed (7). The continuity equation for hydrogen is

$$\frac{dC_{H_s}}{dt} = K_L a_v (C_{H_i} - C_{H_s}) - (-r_H) \tag{CS-13.5}$$

where $1/K_L a_v = (1/k_L a_v) + (1/k_{s_H} a_c)$, and a_v and a_c are the interfacial areas of bubbles and catalyst, respectively, per unit volume of liquid.

Since the flowing gas is essentially pure hydrogen, the equilibrium value C_{H_i} is a constant. It is convenient to solve this equation for the quasi-steady state by setting the left-hand side equal to zero and letting $(-r_H) = r_1 + r_2 + r_5 + r_6$. The resulting equation in terms of $K_d = C_{H_s}/C_{H_i}$ is

$$K_d^{\frac{1}{2}} = \frac{-\left(1 + \dfrac{1}{s_2}\right)(1 - X_B) + \sqrt{\left(1 + \dfrac{1}{s_2}\right)^2 (1 - X_B)^2 + \dfrac{4}{\gamma}\left(\dfrac{1}{\gamma} + \dfrac{Y_{R_1}}{s_5} + \dfrac{Y_{R_2}}{s_6}\right)}}{2\left(\dfrac{1}{\gamma} + \dfrac{Y_{R_1}}{s_5} + \dfrac{Y_{R_2}}{s_6}\right)} \tag{CS-13.6}$$

where $Y_{R_1} = C_{R_1}/C_{B_0}$ and $Y_{R_2} = C_{R_2}/C_{B_0}$, $\gamma = k_1 C_{B_0}/K_L a_v C_{H_i}^{\frac{1}{2}}$, $s_2 = k_1/k_2$, $s_5 = k_1/k_5 C_{H_i}^{\frac{1}{2}}$, and $s_6 = k_1/k_6 C_{H_i}^{\frac{1}{2}}$. When γ is large, the reaction rate is high compared to mass transfer; and large hydrogen concentrations occur making the value of K_d much less than unity. Small values of γ indicate high mass-transfer rates and concentrations of hydrogen at the catalyst surface approaching equilibrium solubility.

Equations for the two yields Y_{R_1} and Y_{R_2} can be obtained by dividing Eqs. CS-13.2 and CS-13.3 by Eq. CS-13.1 to eliminate time and rewriting in terms of X_B, Y_{R_1}, Y_{R_2} and s ratios.

$$\frac{dY_{R_1}}{dX_B} = \frac{(1 - X_B) - \dfrac{Y_{R_1}}{s_3} + \dfrac{Y_{R_2}}{s_4} - \dfrac{Y_{R_1}}{s_5} K_d^{\frac{1}{2}}}{\left(1 + \dfrac{1}{s_2}\right)(1 - X_B)} \tag{CS-13.7}$$

$$\frac{dY_{R_2}}{dX_B} = \frac{\dfrac{1 - X_B}{s_2} + \dfrac{Y_{R_1}}{s_3} - \dfrac{Y_{R_2}}{s_4} - \dfrac{Y_{R_2}}{s_6} K_d^{\frac{1}{2}}}{\left(1 + \dfrac{1}{s_2}\right)(1 - X_B)} \tag{CS-13.8}$$

where $s_3 = k_1/k_3$ and $s_4 = k_1/k_4$. From Eq. CS-13.1 an expression for conversion as a function of time is obtained.

$$\frac{dt}{dX_B} = \frac{1}{(k_1 + k_2)(K_d C_{H_i})^{\frac{1}{2}}(1 - X_B)} \quad \text{(CS-13.9)}$$

The iodine value (IV) can be calculated with respect to conversion and yield assuming 1 mole of iodine consumed per double bond for 18-carbon chains (11).

$$IV = y_{B_0}\left[\frac{2M_1}{M_1}(1 - X_B) + \frac{M_1}{M_2}(Y_R)\right] = y_{B_0}[173.21(1 - X_B) + 86.01 Y_R]$$

$$\text{(CS-13.10)}$$

where X_B is the conversion of diunsaturates, Y_R is the total yield of mono-unsaturates, $Y_{R_1} + Y_{R_2}$, (moles monounsaturate)/(mole of B charged), M_1 is the molecular weight of iodine, M_2 and M_1 are the molecular weights of diunsaturate and monounsaturate, respectively, and y_{B_0} is the mole fraction of B in initial charge.

Equations CS-13.6–CS-13.9 may be solved by a Runge-Kutta routine provided values of the kinetic constants, mass-transfer coefficients and transfer area are known.

The parameter γ provides a useful adjustable parameter for calculating the effects of varying mass-transfer characteristics. The results of such calculations have been presented along with data points from experimental data taken in a 2-liter reactor at low and high rpm (7). Rate constants at 130°C and 60 psig were obtained from analysis of experimental rate plots (5,9). The experimental data at high agitation rates fell essentially on the $\gamma = 0$ lines, which confirms the reasonableness of using kinetic data at high agitation rates as intrinsic values.

Experimental data obtained at a lower rpm (550) conformed more closely to a value between 1.0 and 2.5. To test the usefulness of previous correlations and this model it will be interesting to estimate a value of γ using the experimental kinetic data and correlations given in Chapter 14. All estimates are for 130°C, the temperature of the experiments.

Parameter Estimation

Interfacial Areas

Catalyst: Electronmicrographs of commercial Ruffert-nickel catalyst indicate a variation of particle size. An average value of 150°A was determined.
Based on the experimental loading of 0.07 % nickel

$$m_c = (0.0007)(\rho_L) = (0.0007)(0.91) = 6.37 \times 10^{-4} \text{ g/cm}^3$$

From Eq. 14.25

$$a_c = \frac{6m_c}{D_p \rho_p} = \frac{(6)(6.37 \times 10^{-4})}{(150)(10^{-8})(8.9)} = 286 \text{ cm}^2/\text{cm}^3$$

Gas Bubbles. Sparging velocities and turbulence are rather low in fatty oil hydrogenation and the equation given in Table 14.4 is not applicable. Actual experimental data are preferred. Glycerine adjusted to the same viscosity as cottonseed oil was studied under conditions comparable to the reaction studies at 500 rpm (7). The average bubble diameter was found to be 1 mm and the holdup 3 percent.

Thus

$$a_v = \frac{6H_g}{d_B(1 - H_g)} = \frac{(6)(0.03)}{(0.1)(0.97)} = 1.86 \text{ cm}^2/\text{cm}^3 \text{ of oil}$$

Mass-Transfer Coefficients

Liquid-to-Solid, k_{s_H}. The high surface area of the catalyst makes $k_{s_H} a_c$ high so that the corresponding resistance is low. Thus a rough estimate of k_{s_H} is adequate. Using the equation for very small particles in Table 14.3, a value of 126 cm/sec was estimated based on a liquid diffusivity of 4.73×10^{-5} cm²/sec and a particle diameter of 150A°.

Gas-to-Liquid (k_L°). The equation from Table 14.3 for small bubbles in the size range <4 mm was used and gave a value of $k_L^\circ = 0.012$ cm/sec.

Equilibrium Hydrogen Solubility

Values in cottonseed oil have been correlated with an empirical equation (10) which was extrapolated to 130°C and 60 psig to yield 0.0129 g moles/l.

Calculation of $K_L a_v$

$$\frac{1}{K_L a_v} = \frac{1}{k_L^\circ a_v} + \frac{1}{k_{s_H} a_c} = \frac{1}{(0.012)(1.86)} + \frac{1}{(126)(286)}$$

or

$$K_L a_v = 0.022 \text{ sec}^{-1}$$

Calculation of γ

$$\gamma = \frac{k_1 C_{B_0}}{K_L a_v \sqrt{C_{A_i}}} = \frac{(0.254)(1.45)}{(60)(0.022)\sqrt{0.0129}} = 2.45$$

Comparison of Calculated Values with Experiment

It is evident by reference to Fig. CS-13.2 that the model is directionally correct and, in fact, predicts iodine values, conversion, total monounsaturates and time with reasonable accuracy. Total monounsaturate and *cis*-monounsaturate yields exceed 1.0 because the fresh feed contains *cis*-monounsaturate expressed as $Y_{R_1} = \frac{27}{47} = 0.57$.

The effect of agitation and hydrogen pressure on selectivity and isomerization, described on p. 165, and confirmed by experiment is also reproduced by the model (see Ref. 7). The role of temperature cannot be demonstrated because of the lack of data.

Fig. C5-13.2 is an example of one of many calculations made using various values of k_5 and k_6. Based on the evidence already cited there was reason to assume that k_5 and k_6 are equal. It appears from the results of both these calculations and those in the original reference (7) that this may not be true since the calculated curves are above the experimental values for the *cis*-isomer and below for the *trans*-isomer. Accordingly, a better fit was sought by increasing k_5/k_6 while maintaining the sum of k_5 and k_6 constant. The agreement demonstrated in Fig. C5-13.2 for $k_5/k_6 = 9.29$ is improved but not perfect. The lack of fit of the model to the values of k_5 and k_6 would suggest need for independent evidence to establish the true magnitudes of these constants. The model is obviously not very sensitive to these values as measured by overall performance such as iodine value and reaction time. The latter only differs by 3 min out of 100 for the two extremes of k_5 and k_6.

Further experimental work is needed to define the effect of temperature on the rate constants, the induction period as a function of operating conditions, and determine accurately the effect of catalyst loading. A specific correlation of $K_L a_v$ as a function of power per unit volume for a given reactor configuration is needed for a priori knowledge of γ.

When such data are developed it should be possible to model the entire operating cycle beginning with the heat-up period, provided the parameter γ can be defined for each region of operation. As the intrinsic kinetic data and hydrogen solubilities become more accurately known, it would seem logical in modeling to use γ as an adjustable parameter correlated with operating conditions for a particular reactor system. In this way a useful model for control purposes and operating improvement studies can be evolved. When the model is fully developed it will be possible to determine an optimum value of γ, as well as other operating parameters. Although selectivity and isomerization improve with increasing γ, the time of reaction increases rapidly which creates the classical yield problem. The conversion, time, and iodine value at which the maximum yield of monounsaturates will be obtained can be estimated from plots, such as in Figs. CS-13.2.

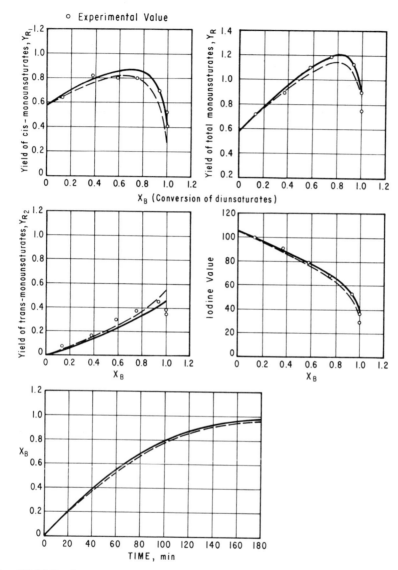

Fig. CS-13.2 Comparison of model with experimental data. (γ = 2.45, k_5 = 1.3, and k_6 = 0.14).

Design Calculations

The complex nature of this system and the present state of the data makes clear the futility of designing or, more accurately, selecting the reactor type and number together with the agitator horsepower solely on the basis of the model just discussed. Plants such as that proposed must be capable of handling several types of feeds and produce varying products depending on marketing conditions.

The actual cycle time has not been reported, but it can be estimated from the model calculations and other considerations. Actual values from an operating plant would be preferable.

load and heat	30 min
evacuate by 3 in. steam ejector to 3 in. Hg, 150 psig steam	20 min
induction time (unsteady-state period)	25 min
reaction time (Fig. CS-13.2) at IV = 65	100 min
analytical	15 min
discharge, cool and filter	90 min
	280 min

An IV of 65 corresponds to 80 % conversion, which is close to the maximum yield of monounsaturates. Higher conversions will produce a lower yield as shown in Fig. CS-13.2.

Highly saturated products require more time and lightly saturated products less. For conservative design assuming different feedstocks and or products each batch, four batches can be made each 24-hr day with plenty of time to allow for proper cleaning and other precautions to prevent intermixing of finished products. Multiple hydrogenators will add flexibility. The design will be based on reactors of 60,000 lb, the capacity of modern tank cars (2). With a required production rate of 9×10^5 to 1×10^6 lb/day, four 60,000 lb reactors will be adequate.

$$(4)(60,000)(4 \text{ batches/day}) = 960,000 \text{ lb/day}$$

Vessel Dimensions

As suggested on p. 645, the residence time of the gas bubbles is increased when tall reactors are used. As the height is increased, however, shaft, bearing, and speed-reducer costs for the agitator assembly increases. Mixer manufacturers should be consulted for detailed recommendations. A ratio of

$L/D \approx 2$ is commonly used in stirred gas contactors, and a shaft with steady bearing can be specified as shown in Fig. CS-13.3 without excessive cost. Use a freeboard of 15–20% of the volume of the liquid for hydrogen gas space (2). Design for possible upper limit of temperature at 200°C (1) for which $\rho_L = 0.8$ (14).

$$\text{Volume} = \frac{(60,000)}{(0.80)(62.4)} = 1201.9 \text{ ft}^3$$

Add 5% to approximate volume occupied by coil and correction for head.

$$\frac{\pi D^2}{4} L = \frac{\pi D^3}{2} = (1201.9)(1.05)$$

or

$$D = 9.3 \text{ ft}$$

Use standard 112 in. elliptical head with 112.2 cu ft capacity.

Impeller Selection

Three impellers should be used, with the first located $\frac{1}{6}D$ above the vessel bottom as given in Table 14.5. The most intense scale of agitation will be governed by heat transfer during cooling of the batch to 90°C, for which use an impeller-to-vessel diameter ratio of $\frac{1}{3}$.

$$D_I = \tfrac{112}{3} \approx 38 \text{ in.}$$

We will specify this diameter. The nearest manufacturer's standard which may vary with manufacturer, should be used. First impeller position will be approximately 19 in. above bottom head, which for that head corresponds to 9 in. below tangent line. A sparge ring shall be located just below this first impeller with perforations directed downward (see Table 14.5). The second impeller should be located half-way between the first and the top impeller ($\approx$ one tank diameter). The two lower impellers should be disk-type, flat-blade turbines with six blades. Agitation for good heat transfer and turn-over but not high shear is wanted in this case. Specify a tip speed from Table 8.10 of 1000 ft/min.

$$\text{rpm} = \frac{1000}{(\pi)(38/12)} = 100.5 \text{ rpm}$$

Lower speeds will be possible during hydrogenation after the pressure has stabilized. Specify a variable-speed drive for greater flexibility in product quality and energy savings. Hydraulic drives are attractive, for they are explosion proof and can be easily designed for automatic overload protection.

A third impeller of pitched-blade type the position of which is adjustable should be located just below the operating liquid surface (6–12 in.) to assist in aspirating hydrogen from the vapor space into the oil during a major portion of the cycle when little hydrogen is being added. Four standard baffles ($0.1D = 11$ in.) will be specified (see Fig. 8.15) set out $\frac{1}{6}$ of baffle width from wall (Table 14.5). The top of the baffles should be about 12 in. below the operating liquid surface (12) so that some vortexing will occur and aid in inducing gas from the freeboard. Because of the lower power number for a pitched-blade impeller (see Fig. 8.7), it will be specified with a larger diameter in order to balance the shaft load.

$$(N_p D_1^5)_2 = (N_p D_1^5)_1$$

or

$$(D_1)_2 = \left(\frac{N_{p_1}}{N_{p_2}}\right)^{\frac{1}{5}} (D_1)_1 = \left(\frac{5}{2.1}\right)^{\frac{1}{5}} (38) = 45 \text{ in.}$$

N_{p_2} was obtained by ratioing curves 3 and 6 of Fig. 8.7 with curve 1 so that pitch-blade impeller would have the same blade width as the others.

Agitator Horsepower

Curves 1 and 6 of Fig. 8.7 apply. Calculations are based on 90°C, the temperature of the cooled oil prior to discharge. Since the agitator must be operated with no gas during cool down, motor selection will be based on all liquid.

$$N_{Re} = \frac{\rho_L N D_1^2}{\mu_L} = \frac{(0.87)(100)[(38)(2.54)]^2}{(60)(9.2 \times 10^{-2})} = 146{,}834$$

Property values are from Ref. 7, 13, and 14. Use power consumption for single pitch-blade and two single flat-blade turbines times 0.9 (see p. 348)

$$P_a = \left[\frac{(0.89)(62.4)\left(\frac{100}{60}\right)^3}{32.17}\right]\left[\left(\frac{38}{12}\right)^5 (2)(5) + \left(\frac{45}{12}\right)^5 (2.1)\right] 0.9$$

$$= 34106 \text{ ft lb}_f/\text{lb}_m \quad \text{or} \quad \frac{34106}{550} = 62.0 \text{ hp}$$

Add 10% for driver loss and 0.5 hp seal loss. Specify nearest standard explosion-proof motor which is 75 hp. The power input at 130°C with gas flowing was calculated to be 55.5 hp based on u_s of 0.01 ft/sec and Eq. 14.8 for first impeller and 0.7 and 0.9 factors for no-gassed power for top and middle impellers, respectively. Another common approach to a design of this type is to use an accepted hp/1000 gal. and calculate the required impeller diameters.

Heat Exchange

A heating coil will be more economical since jacketed vessels in this configuration are not standard. Heating the charge will be more efficiently accomplished by passing through an external heat exchanger. The coil will be used to cool the reaction mixture. The maximum possible heat load would occur for the case of total mass-transfer control. This could happen after the reaction pressure had been reached due to a temperature excursion. Heat of reaction data are sparse. A reported value for total hydrogenation of olein (15) is equivalent to 29.4 kcal/g mole of H_2 reacted. This compares well with 30.4 kcal/g mole for octadecene $\rightarrow$ octadecane. The maximum cooling load occurs just after the induction period and at higher temperature could correspond to the maximum rate of hydrogen transfer.

$$K_L a_v C_{H_i} = (0.022)(0.0129)(60) = 0.017 \text{ g moles/(l.)(min)}$$

$$\text{Volume of oil in vessel} = \frac{(60,000)(28.34 \text{ l./ft}^3)}{(0.84)(62.4)} = 32395 \text{ liters}$$

$$\text{Cooling load} = (0.017)(29,400)(60)(32395)$$

$$= 971.46 \times 10^6 \text{ cal/hr or } 3.855 \times 10^6 \text{ BTU/hr}$$

$$\rho_L = 0.84 \text{ at } 130°C \text{ (13)}$$

With reference to the discussion on p. 378[l], it should be clear that better control can be attained with a steep heat removal curve which means a modest ΔT between the reacting mixture and the cooling medium. This is particularly true for oil-H_2 which does exhibit a rapidly rising heat-generation curve because of the strong mass-transfer effects with rising temperature that are not very temperature sensitive. Unfortunately we have kinetic data at but one temperature and can only speak qualitatively. Therefore, with a 130°C (265°F) operating temperature select an average cooling temperature of 200°F as an estimate of a proper value. This will require a closed pressure-condensate recirculating system with an external heat exchanger for removing the heat from the condensate. One major manufacturer of hydrogenation reactor systems of this type offers a patented process involving vaporizing water in the coil which affords excellent control of temperature and thereby product quality (16).

Use a 20°F temperature rise for the water.

$$3.722 \times 10^6/20 = 186100 \text{ lb/hr of water or 372 gpm}$$

Design for approximately 8 ft/sec to assure high coefficient. Schedule 40 carbon steel pipe produces a velocity of 9.33 ft/sec and a head loss per 100 ft of straight pipe of 13.4 ft-lb$_f$/lb$_m$ (5.8 psi), which is a little high. Although 5 in. pipe is not as widely used in process piping, it is manufactured and will be specified (velocity = 5.96 ft/sec and head loss = 4.5 ft lb_f/lb_m or ≈ 2 psi/100 ft). Even in turbulent flow at this curvature some additional loss will occur due to the helical shape, but the total loss will not be over 10 or 12 psi in 500 ft of pipe.

Coil Configuration. The heat-transfer equation for the agitated side is based on $D_c/D = 0.7$. Hence coil diameter should be in this range, $(0.7)(112) \approx 80$ in.

Inside Heat-Transfer Coefficient. From Appendix F,

$$h_i = (2150)\left(\frac{0.62}{5.047}\right)^{0.2} = 1413.$$

This will be increased because of the coiled shape (see p. 361[1]), 1413[1 + (3.5)(5/80)] = 1722.

Heat-Transfer Coefficient on Agitated Side. Although sparging increases the heat transfer coefficient (see Eq. 14.10), we will base design on conservative case where gas is mainly being drawn from freeboard. From Table 8.6 and properties based on Ref. 13 and 14 (see also p. 357[1]).

Basis. 130°C (266 F); one impeller, no gas

$$h = \frac{\lambda_f}{d_{ct}}(0.17)\left(\frac{\rho_f N D_I^2}{\mu_f}\right)^{0.67}\left(\frac{c_p \mu_f}{\lambda_f}\right)^{0.37}\left(\frac{\mu_b}{\mu_w}\right)^{0.54}\left(\frac{D_I}{D}\right)^{0.1}\left(\frac{d_{ct}}{D}\right)^{0.5}$$

$$= \frac{(12)(0.089)}{(5.563)}(0.17)\left[\frac{(0.84)(62.4)(100)(60)(38/12)^2}{(4.5)(2.42)}\right]^{0.67}$$

$$\times \left[\frac{(0.57)(4.5)(2.42)}{0.089}\right]^{0.37}\left(\frac{4.5}{8.2}\right)^{0.54}\left(\frac{38}{112}\right)^{0.1}\left(\frac{5.563}{112}\right)^{0.5}$$

$$= 105 \text{ BTU/hr ft}^2 \text{ F or } (105)[(3)(0.9)]^{0.22} = 131 \text{ for 3 impellers.}$$
$$h_g = (131)(55.5/60.6)^{0.25} = 128$$

Using fouling factor of 0.004, 0.001 water side and 0.003 oil side (TEMA),

$$U = 1 \bigg/ \left[\left(\frac{1}{128}\right) + 0.004 + \left(\frac{1}{1722}\right)\right] = 80.7$$

Coil Length (area per foot = 1.456 sq ft for 5 in. Schedule 40)

$$\frac{3.855 \times 10^6}{(66)(81)(1.456)} = 495 \text{ ft}$$

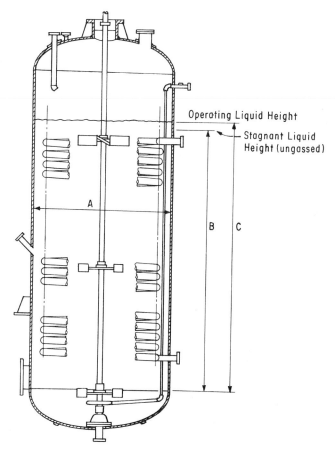

Operating Liquid Height

Stagnant Liquid
Height (ungassed)

Fig. CS-13.3 Final design of hydrogenator (similar to hydrogenators supplied by Votator Division of Chemetron Corp., Louisville, Ky.).
Summary of Calculated Results:

Operating temperature   266°F (130°C)
Operating pressure   60 psig
Nominal capacity   60000 lb
Operating capacity   1105.2 ft³ (8267 gal.)
Capacity of head   106.4 ft³ (dished portion)
Volume occupied by coil and agitator shaft: 96.46 ft³
Volume occupied by freeboard: 221 ft³ to tangent line
Dimensions: A = 112 in., B = 16 ft, C = 16.5 ft (3% holdup), and tangent-to-tangent = 19.73 ft.

Based on cooling the ungassed oil from $130°C$ to $90°C$ in $\frac{1}{2}$ hour, a similar calculation with $U = 59$ yields 512 ft of coil which will be specified.

The coils should be spaced to distribute over the length of the liquid zone beginning at $(0.15)(112) = 17$ in. from the vessel bottom (see Table 8.6). Since the baffles consume $2(11) = 22$ in. of the inside diameter and the coil was tentatively specified as 80 in., it may be advantageous to increase the diameter of the coil to $112-122 = 90$ in. so that it can be supported by the baffles.

Figure CS-13.3 is a pictorial of a reactor of the type designed and constructed by a company which specializes in reactors of this type. On the drawing are summarized the results of these calculations.

REFERENCES

1. L. F. Albright, *Chem. Eng.* (*N.Y.*), p. 197 (Sept. 11, 1967).
2. L. F. Albright, *Chem. Eng.* (*N.Y.*), p. 249 (Oct. 9, 1967).
3. *Food Eng.*, p. 140 (Mar., 1954).
4. R. R. Allen, private communication, April 9, 1973.
5. K. Hashimoto, K. Muroyama, and S. Nagata, *J. Am. Oil Chem. Soc.*, **48**, 291 (1971).
6. L. F. Albright and J. Wisniak, *J. Am. Oil Chem. Soc.*, **39**, 14 (1962).
7. K. Hashimoto, M. Teramoto, and S. Nagata, *J. Chem. Eng. Japan*, **4**, 150 (1971).
8. R. R. Allen and A. A. Kiess, *J. Am. Oil Chem. Soc.*, **33**, 355 (1956).
9. I. A. Eldib and L. F. Albright, *Ind. Eng. Chem.*, **49**, 825 (1957).
10. J. Wisniak and L. F. Albright, *Ind. Eng. Chem.*, **53**, 375 (1961).
11. L. F. Albright, *J. Am. Oil Chem. Soc.*, **42**, 250 (1965).
12. Bulletin PTE-2, Philadelphia Mixers Corp., King of Prussia, Pa., 1970.
13. A. E. Bailey, *Cottonseed and Cottonseed Products*, Interscience, New York, 1948.
14. *Kirk–Othmer Encyclopedia*, Vol. 8, Wiley-Interscience, New York, 1965, p. 776.
15. R. N. Shreve, *Chemical Process Industries*, 3rd ed., McGraw-Hill, New York, 1956.
16. Votator Division Chemetron Corp., Louisville, Ky., U.S. Patent 3732266, 1974.

CASE STUDY 114

Hydrodesulfurization

THIS PROCESS of decomposition of a variety of organic sulfur compounds in petroleum involves not only complex chemistry but also transport resistances and distribution problems. Expensive pilot-plant work must often be done to assure valid design decisions. Scale-up of such studies is widely practiced. This case study simulates a scale-up required on a catalyst for which limited data have been obtained. Previous more detailed studies on another catalyst are used to aid in this bold extrapolation.

Problem Statement

Pilot-plant data shown in Fig. 14.17, p. 710, are for 50 % hydrodesulfurnization of vacuum-flashed distillate at 375°C and 740 psi hydrogen pressure and a 2300 scf/bbl hydrogen-to-oil ratio. The intercept value of $\tau_{0.5}$ (reaction half-time) corresponds to that for an autoclave experiment. A new catalyst with longer life but the same dimensions has been developed and tested in the autoclave at the same conditions and a value of $\tau_{0.5} = 0.15$ was observed. Assuming a quick answer without pilot-plant work is needed, size a reactor for this new catalyst using a feed rate of 1000 bbl/hr. The 21.9° API oil contains 1.9 wt % sulfur. A flash calculation using the ASTM distillation was made to assure that most of the feed remains as a liquid.

Catalyst

The catalyst is cobalt oxide-molybdenum oxide-alumina, $\frac{1}{8}$ in. $\times \frac{1}{8}$ in. cylindrical pellets, $\rho_b = 50$ lb/ft^3, and bed void fraction $= 0.4$. Temperatures above 1150°F should be avoided. The catalyst reaches its full activity after a period of use with a sulfur containing feed which produces the active sulfided form.

Chemistry

Although precise mechanistic details are lacking it is generally agreed that the sulfur compound adsorbs on molybdenum atoms which are part of a MoS_2 network. In the case of ring compounds such as thiophene the adsorbed species is thought to first rupture at the CS bound followed by hydrogenation of the straight chain compound and the adsorbed sulfur to produce hydrocarbons of various degrees of saturation and H_2S (1). One of the surface steps is usually thought to be rate controlling. Typical overall reactions in order of decreasing ease of hydrogen analysis are

$$\text{(mercaptan) } RSH + H_2 \longrightarrow RH + H_2S$$
$$\text{(disulfide) } (RS)_2 + 2H_2 \longrightarrow 2RH + H_2S$$
$$\text{(sulfide) } R_2S + 2H_2 \longrightarrow 2RH + H_2S$$
$$\text{(thiophene) } C_4H_4S + 4H_2 \longrightarrow C_4H_{10} + H_2S$$

Aromatic thiophenes are the most difficult to desulfurize.

Thermodynamics

Both desulfurization and the incidental hydrogenation of other compounds present in the feed are exothermic reactions. Temperature rise is normally based on hydrogen consumption data from pilot plant or commercial plant and a reaction heat per unit of hydrogen consumed (50–100 BTU/scf of H_2 consumed). This quantity varies with the aromatics and olefinic composition of the feed (4). For purposes of this scale-up assume equal temperature profiles for each case.

The equilibrium constants for sulfur compounds below 600°C are all large and the reactions are not equilibrium limited.

Design Calculations

Solve for $(u_s)_L$

Basis. $G_L = 5000$ lb/(hr)(ft^2) (see p. 700[1])

$$\rho_L \text{ at } 375°C = 43.6 \text{ lb/ft}^3 \qquad (API \ Data \ Book)$$

$$(u_s)_L = \frac{5000}{43.6} = 114.7 \text{ ft/hr}$$

Reactor Efficiency (Original Catalyst)
From Fig. 14.17, $\tau_{0.5} = 0.2$ hr and ideal $\tau_{0.5} = 0.12$.

$$E_B = \frac{0.12}{0.2} \times 100 = 60\%$$

Assuming same efficiency for new catalyst

$$\text{bed height} = \tau_{0.5}(u_s)_L = \left(\frac{0.15}{0.6}\right)(114.7) = 28.7 \text{ ft}$$

$$\text{diameter} = \left[\frac{(1000)(42)(57.55)(4)}{(7.48)(5000)(\pi)}\right]^{\frac{1}{2}} = 9.07 \text{ ft}$$

This reactor should be provided with a distributor as discussed on p. 700[l].
Check pressure drop (Eq. 14.65) $\varepsilon = 0.4$

$$\log_{10}\left(\frac{\Delta P_{TP}}{\Delta P_G + P_L}\right) = \frac{0.7}{[\log_{10}(\chi/1.2)]^2 + 1.0}$$

$$\frac{\Delta P}{\Delta Z} = \left(\frac{f_k \mu u_s}{D_p^2 g_c}\right)\frac{(1-\varepsilon)^2}{\varepsilon^3}$$

$$f_k = 150 + 4.2\left[\frac{D_p u_s \rho}{\mu(1-\varepsilon)}\right]^{\frac{5}{6}}$$

Liquid. $\mu = 0.5$ cp (*API Data Book*)

$$f_k = 150 + 4.2\left[\frac{(0.125)(5000)}{(12)(0.5)(2.42)(0.6)}\right]^{\frac{5}{6}} = 297.8$$

$$\frac{\Delta P}{\Delta Z} = \frac{(297.8)(0.5)(2.42)(5000/43.6)(0.6)^2}{(0.125/12)^2(32.17)(3600)^2(0.4)^3} = 5.13 \text{ psf/ft}$$

Gas. $\mu = 0.017$ $\rho_g = 0.12$ lb/ft^3

$$G = \frac{(2.3 \times 10^6)(2)}{(379)(\pi/4)(9)^2} = 190.8 \text{ lb/hr ft}^2$$

$$u_s = \frac{190.8}{0.12} = 1590 \text{ ft/hr}$$

$$f_k = 150 + 4.2\left[\frac{(0.125)(190.8)}{(12)(0.017)(2.42)(0.6)}\right]^{\frac{5}{6}} = 312.7$$

$$\frac{\Delta P}{\Delta Z} = \frac{(312.7)(0.017)(2.42)(1590)(0.6)^2}{(0.125/12)^2(32.17)(3600)^2(0.4)^3} = 2.55 \text{ psf/ft}$$

Total 7.68 psf/ft

$$\log_{10}\left(\frac{\Delta P_{TP}}{7.68}\right) = \frac{0.7}{\left\{\log_{10}\left[\left(\frac{5.13}{2.55}\right)^{\frac{1}{2}}\Big/1.2\right]\right\}^2 + 1.0}$$

$\Delta P_{TP} = 38.17$ psf/ft or 0.27 psi/ft

This is a low ΔP (7.75 psi for entire bed) but as coking occurs the porosity of the bed will decline. Consider, for example, going from $\varepsilon = 0.4$ to $\varepsilon = 0.2$. ΔP_{TP} would increase to 3.4 psi/ft and this represents a ΔP of 97.6 psi for the entire bed.

Future expansion can be usefully studied on the basis of decreasing particle size which reduces the required reactor volume for a catalyst of any porosity. The reduction is the most dramatic for heavier stocks because diffusion limitations are greater (2). By reducing catalyst size a heavier stock can be successfully handled in a reactor designed for a lighter feed, or deeper sulfur removal of the same light feed can be accomplished. In most cases $\frac{1}{16}$ in. is the smallest practical size unless bed deposits can be successfully prevented. In considering expansion possibilities, the designer should not neglect the extra compressor horsepower required for the higher ΔPs produced with smaller catalyst.

REFERENCES

1. P. Kieran and C. Kemball, *J. Catalysis*, **4**, 394 (1965).
2 J. W. Scott and A. G. Bridge, *Adv. Chem. Ser.*, **103**, 113 (1971).
3. A. Bondi, *Chemtech*, p. 185 (March, 1971).
4. S. C. Schuman and H. Shalit, *Catal. Rev.*, **4**, 245 (1970).

APPENDIXES

APPENDIX A

Frequently Used Constants and Factors

Selected Basic and Defined Constants

Constant	Symbol	Value
Absolute temperature of triple point of water	T_{tp}	273.16°K, 491.69°R
Acceleration of gravity	g	980.665 cm sec^{-2}, 32.174 ft sec^{-2}
Avogadro number	N_o	6.02252 × 10^{23} molecules mole^{-1}
Boltzmann constant	$k = R/N_o$	1.38054 × 10^{-6} erg °K molecule^{-1}
Calorie (thermochemical)	cal	4.184 abs joules, 3.9657 × 10^{-3} BTU
Calorie (international steam)	cal$_{IT}$	4.1868 abs joules, 3.9683 × 10^{-3} BTU
Gas constant	R	8.31433 joules (g mole)$^{-1}$ °K^{-1}
		82.056 cm^3 atm (g mole)$^{-1}$ °K^{-1}
		10.731 (psia) ft^3 (lb mole)$^{-1}$ °R^{-1}
		0.7302 (atm) ft^3 (lb mole)$^{-1}$ °R^{-1}
	R'	1.9817 cal (g mole)$^{-1}$ °K^{-1}
Inch	in.	2.54 cm
Planck constant	h	6.6256 × 10^{-27} erg sec molecule^{-1}
Pound, avoirdupois	lb	453.59237 g
Pressure-volume product for 1 mole of gas at 0°C and zero pressure	$(PV)_{0°C}^{P=0}$	2271.06 joules/g-mole
		22413.6 cm^3 atm (g mole)$^{-1}$
		359.03 ft^3 atm (lb mole)$^{-1}$
Standard atmosphere	atm	1,013,250 dynes cm^{-2}
		14.696 psia, 1.0332 kg/cm^2
		760 mm Hg, 29.921 in. Hg, 33.9 ft H$_2$O
Degree Celsius	°C	T°K − 273.15
Degree Fahrenheit	°F	T°R − 459.67
Degree Rankine	°R	$(\frac{5}{9})$°K

Note. These constants are given as the most accurate values. In normal engineering work rounded-off values would be used.

SELECTED CONVERSION FACTORS. (Multiply numerical value of left-hand column by values given to obtain units indicated)

Viscosity	poise, g/cm sec	lb_m/ft sec	lb_m/ft hr	kg_m/m sec	kg_m/m hr
Centipoise	0.01	6.72×10^{-4}	2.419	0.001	3.6

Energy	BTU	Calories	ft-lb_f	hp-hr	Joules abs	KW-hr
BTU	1	252.16	778.16	3.9301×10^{-4}	1005	2.9307×10^{-4}
calories	3.9657×10^{-3}	1	3.086	1.5586×10^{-6}	4.184	1.1622×10^{-6}
ft-lb_f	1.2851×10^{-3}	0.32405	1	5.0505×10^{-7}	1.3558	3.7662×10^{-7}
hp-hr	2544.5	6.4162×10^{5}	1.9800×10^{6}	1	2.6845×10^{6}	0.74570
Joules (abs)	9.4783×10^{-4}	0.23901	0.73756	3.7251×10^{-7}	1	2.7778×10^{-7}
KW-hr (abs)	3412.2	8.6042×10^{5}	2.6552×10^{6}	1.3410	3.6000×10^{6}	1

Power	BTU/hr	cal/sec	ft-lb_f/min	hp	Watts
BTU/hr	1	0.70045	12.969	3.9301×10^{-4}	0.29307
cal/sec	14.277	1	185.16	5.6108×10^{-3}	4.1840
ft-lb_f/min	0.77105	5.4008×10^{-3}	1	3.0303×10^{-5}	0.022597
hp	2544.5	178.23	33,000	1	745.7
Watts	3.4122	0.23901	44.254	1.3410×10^{-3}	1

APPENDIX B

Vessel Design and Costs

VESSEL COSTS are an important element in reactor design decisions. The most reliable and rapid method requires the assistance of a representative of a vessel fabricator. These specialists can usually supply cost-per-unit-weight data for the particular type of vessel and material of construction together with separate nozzle costs usually as a cost per unit diameter. Internals such as support plates are best handled as separate items using a cost/mass value for the type of internal. Rapidly changing economic conditions and the wide variety of reactor types make any listing of such costs unwise. Instead some charts and tables are assembled here which will facilitate estimating weights and calculating vessel thickness.

CODES

In the U.S.A. the American Society of Mechanical Engineers has established a code for the design and fabrication of pressure vessels. Similar organizations in Europe have established codes. All such codes give minimum standards. Normally vessels as important as reactors are designed to comply not only with a code but also with supplemental specifications considered important for the particular service. These can include special impact test requirements to assure against brittle fracture, heat-treating specifications for steel in severe service such as high hydrogen partial pressures, and unique rules for pressure testing (1).

Economical, safe, and durable pressure vessels require the design abilities of an expert; but the codes, although not handbooks, can serve as useful guides to the process designer wishing to make approximate cost estimates for use in design decision. For this purpose design equations from the ASME Code are summarized. Any critical cost study that could involve small cost differences which become significant over a span of years, should be made using actual quotations with the advice of a vessel expert so that the most economic design will be obtained.

DESIGN EQUATIONS

Equations for calculating vessel thicknesses are assembled in Table B.1 for vessels under internal pressure. For vessels under external pressure, such as a vessel operating under vacuum or an internal shell operating at a lower pressure than that in the external shell, see special charts in ASME Code.

The equations given in Table B.1 and the allowable stresses in Table B.2 are from the most widely used Section VIII, Division 1 of the ASME Code. Division 2 standards require more rigid fabrication and design procedures which allow some savings in vessel weight. In the case of large vessels fabricators may determine that Division 2 standards offer savings.

STANDARD VESSEL SIZES

Greater size standardization in the pressure vessel industry would no doubt reduce fabricating costs. It has been suggested that if customers of vessel fabricators would standardize on 1-ft diameter increments for vessels 4–8 ft in diameter and 2-ft increments from 8 through 14 ft that four standard plate widths could be established (4). The result would be savings in both steel mill costs and handling and storage costs at the fabricator's plant. No such general practice exists today, but in dealing with major vessels the designer should consult with vessel fabricators on availability and costs for various options. Vessel heads can be purchased in an increasing variety of sizes and place very few limits on the diameter selected up to ≈ 23 ft. Above this value heads must be built-up from sections of plate.

Standard plate thicknesses for carbon steel pressure vessels are in $\frac{1}{16}$ in. increments between $\frac{3}{16}$ in. and 1 in., $\frac{1}{8}$ in. increments between 1 in. and 2 in., $\frac{1}{4}$ in. between 2 in. and 4 in., $\frac{1}{2}$ in. between 4 in. and 8 in. and 1 in. in larger thicknesses. Most low alloy and all high alloy steel is so much more costly that it is advantageous to order to the calculated decimal thickness especially for large vessels requiring substantial amounts of steel plate.

Table B.1 Equations for Calculating Pressure Vessel Thickness

Shape	Equations (Internal Pressure)	Conditions
Cylindrical	$t = \dfrac{PR}{SE - 0.6P}$	$t \le R/2$
		$P \le 0.385\,SE$
	$t = R\left(\dfrac{SE + P}{SE - P}\right)^{\frac{1}{2}} - R$	$t > R/2$
		$P > 0.385\,SE$
Spherical	$t = \dfrac{PR}{2SE - 0.2P}$	$t \le 0.356\,R$
		$P \le 0.665\,SE$
	$t = R\left(\dfrac{2SE + 2P}{2SE - P}\right)^{\frac{1}{2}} - R$	$t > 0.356\,R$
		$P > 0.665\,SE$
Ellipsoidal head	$t = \dfrac{PD}{2SE - 0.2P}$	0.5 (minor axis) $= 0.25D$
Torispherical head	$t = \dfrac{0.885\,PL}{SE - 0.1P}$	Knuckle radius, r, is 6% of inside crown radius
Hemispherical	Same as spherical vessel with R being inside spherical radius in inches.	
Conical	$t = \dfrac{PD}{2 \cos \alpha\,(SE - 0.6P)}$	$\alpha \le 30°$

From: Section VIII, Div. 1, *ASME Boiler and Pressure Vessel Code*, American Society Mechanical Engineers, New York, 1974.

Special Nomenclature. To ensure continuity nomenclature of the ASME Code has been retained; t is the minimum thickness of shell or head exclusive of corrosion allowance, inches, P is the design pressure, maximum difference in pressure between inside and outside of shell, psi; R is the inside radius of shell before corrosion allowance is added, inches; S is the maximum allowable stress value, psi (see Code for complete data; some values given for typical steels in B-2); E is the joint efficiency (for double-welded butt joints = 1.0 if fully radiographed, 0.85 if spot examined, and 0.7 if not radiographed (see Code for details); D is the diameter of head skirt or inside length of major axis of an ellipsoidal head or inside diameter of cone head; L is the inside crown radius, α = one-half of included angle of the cone at centerline of head, and r is the knuckle radius.

Note. Thickness of lining for clad reactors not allowed in determining allowable pressure for applied linings. For integrally clad linings credit is allowable for certain steels (see Code).

Table B.2 Selected Plate Steels for Reactor Fabrication

(handwritten: lb_f = 1000 kips)

ASTM Designation	Nominal Composition	Allowable Stress, Kips/in². (ASME Code—Sect. VIII, Div. 1, 1974) Temperature, °F											
		−20 to 650	700	750	800	850	900	950	1000	1050	1100	1150	1200
Carbon steels													
A-285 Gr. C	C-Si	13.7	13.2	12.0	10.2	8.3	6.5	(intermediate tensile strength) See Note 1					
A-515 Gr. 70	C-Si }	17.5	16.6	14.7	12.0	9.2	6.5	4.5	2.5	See Note 1			
A-516 Gr. 70	C-Si }	(for intermediate and high temperature service)											
		(for lower temperature service requiring notch toughness)											
Low-alloy steels for resistance to H_2 and H_2S (see Fig. B.1)													
A204 Gr. A	C-0.5Mo	16.2	16.2	16.2	15.6	14.4	12.5	10.0	6.2	See Note 2			
Gr. B	C-0.5Mo	17.5	17.5	17.5	16.9	15.0	13.0	10.0	6.2	See Note 2			
Gr. C	C-0.5Mo	18.7	18.7	18.7	18.0	15.9	13.0	10.0	6.2	See Note 2			
A387 Gr.12, Cl.1, 1Cr-0.5Mo		13.7	13.7	13.7	13.7	13.6	13.0	11.0	7.5	5.0	2.8	1.5	1.0
A387 Gr.22, Cl.2, 2.5Cr-1Mo		See Code	17.1	17.0	16.8	16.3	14.7	11.0	7.8	5.8	4.2	3.0	1.6
Cl.1 2.25Cr-1Mo		15.0	15.0	15.0	15.0	14.4	13.1	11.0	7.8	5.8	4.2	3.0	1.6
A387 Gr.5, Cl.1 5Cr-0.5Mo		See Code	13.4	13.1	12.8	12.0	10.3	7.6	5.6	4.1	3.0	2.0	1.3
High tensile steels for heavy-wall reactors													
A302 Gr.B	Mn-0.5Mo	20.0	20.0	20.0	19.1	16.8	13.2	10.0	6.2				
Gr.C	Mn-0.5Mo-Ni												
A533 Gr.A, Cl.1	C-Mn-Mo	20.0	20.0	20.0	19.1	16.8	13.2	10.0	6.2				
Gr.B, Cl.1	C-Mn-Mo., 0.4–0.7 Ni }												
Gr.C, Cl.1	C-Mn-Mo., 0.7–1.0 Ni }	20.0	20.0	20.0	19.1					(Quenched and tempered all grades)			

A542, Cl.1 (Same analysis as A387 Gr.22 Cl.2 but quenched and tempered. Increases tensile strength from 75 to 105. Allowable stresses listed only under Div. 2 Rules)

High-alloy steels for cladding and internal shells for corrosion resistance

A240 304	18Cr-8Ni	18.7 @ −20°F 11.2 @ 650°F	11.0	10.8	10.5	10.3	10.1	9.9	9.7	9.5	8.8	7.7	6.0 See Notes 4 and 5 (4.7 @ 1250°F to 1.4 @ 1500°F)
304L	18Cr-8Ni	15.6 @ −20°F 9.5 @ 650°F	9.3	9.2 See Note 4	9.0	(Type 304 and 364L for general corrosion resistance)							
316 (See Note 6)	16Cr-12N-2Mo	18.7 @ −20°F 11.5 @ 650°F	11.3	11.1	11.0	10.9	10.8	10.7	10.6	10.5	10.3	9.3	7.4 See Notes 4 and 5 (5.4 @ 1250°F to 1.2 @ 1500°F)
316L (See Note 3)	16Cr-12Ni-2Mo	15.6 @ −20°F 9.1 @ 650°F	8.9	8.7	8.6	8.4	(Type 316 and 316L for more severe corrosion resistance) See Note 4						
321	18Cr-10Ni-Ti	18.7 @ −20°F 11.1 @ 650°F	10.9	10.8	10.7	10.6	10.5	10.4	9.2	6.9	5.0	3.6 See Notes 4 and 5 (2.5 @ 1250°F to 0.3 @ 1500°F)	
		(Stabilized to prevent carbide precipitation in 800–1500°F range)											
347	18Cr-10Ni-Cb	18.7 @ −20°F 13.1 @ 650°F	12.9	12.8	12.7	12.6	12.5	12.5	11.9	9.1	6.1	4.4 See Notes 4 and 5 (3.3 @ 1250°F to 0.7 @ 1500°F)	
		(similar to 321)											
405	12Cr-Al	15 @ −20°F 12.2 @ 650°F	12.0	11.6	11.1	10.4 See Note 7	9.6	8.4	4.0	(better weldability and ductility than 410; good for reactor tubing, SA268-TP405)			

Table B.2 (*Continued*)

| ASTM Designation | Nominal Composition | Allowable Stress, Kips/in². (ASME Code—Sect. VIII, Div. 1, 1974) Temperature, °F | | | | | | | | | | | |
		−20 to 650	700	750	800	850	900	950	1000	1050	1100	1150	1200
410S	13Cr	15.0 @ −20°F 12.2 @ 650°F	12.0	11.6	11.1	10.4	9.6	8.4	6.4	4.4	2.9	1.7	1.0
			(used in high-pressure reactor tubing, SA 268–410) See Note 8										
430	17Cr	16.2 @ −20°F 13.2 @ 650°F	13.1	12.6	12.0	11.2	10.4	9.2	6.5	4.5	3.2	2.4	1.7
			(resistant to highly oxidizing atmospheres) See Note 7										

1. Temperatures above 800°F for prolonged periods may cause graphitization of carbide phase. Use killed steel above 850°F for A-285 GR.C with not less than 0.1% residual silicon.

2. Temperatures above 875°F for prolonged periods may cause graphitization of carbine phase.

3. *L* types are specified if welding is done without solution heat treatment.

4. Higher stress values for Type 304 and 304L are listed in the Code when slight deformation is acceptable.

5. These stresses apply above 1000°F if C > 0.04% and if heat treatment at 1900°F is applied followed by rapid quench (see Code).

6. Type 317, which is even more corrosion resistant than 316, has same stress values. Type 317L has allowable stress values approximately 25% higher than 316L (see Code).

7. Develops embrittlement at room temperature when used above 700°F.

8. Slightly higher stresses for plate allowed for Type 410.

Choice of steels listed based largely on J. F. Lancaster, *Hydrocarbon Process.*, **49** (6) 84 (1970). Refer to this reference for comparison with European Codes. Allowable stress from *ASME Boiler and Pressure Vessel Code*, Section VIII, Div. 1, American Society of Mechanical Engineers, New York, 1974.

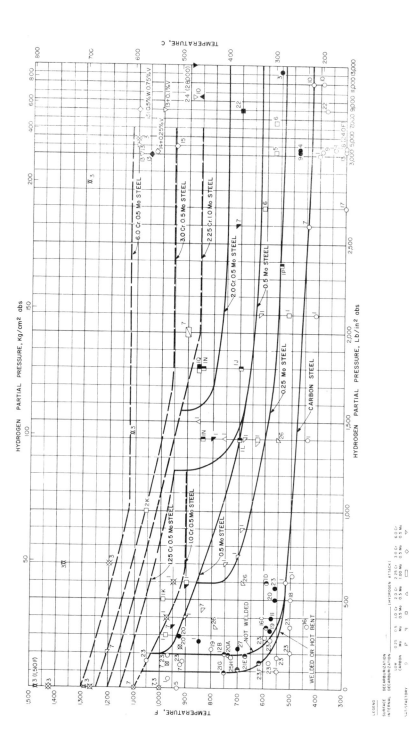

Fig. B.1 Operating limits for steels in refinery service [*Notes*: 1. Austenitic stainless steels are satisfactory at all temperatures. 2. Effect of trace alloying elements is shown in Fig. 2 of original reference. 3. See references and footnotes in original reference section. Reproduced by permission: API Pub. 941, American Petroleum Institute, Washington, D.C., 1970 (original copyright by G. A. Nelson).

CORROSION ALLOWANCE

Although practices vary, on the average a material is selected that will not corrode more than 0.010 to 0.015 in./yr (10–15 mpy or 0.25–0.36 mm/yr). For a vessel life of 10 yr this approximates a corrosion allowance of $\frac{1}{8}$ in. In some systems it is cheaper to accept more corrosion if uniform corrosion is expected than to specify more expensive alloys. Usually when the indicated allowance exceeds $\frac{1}{4}$ in., a more resistant metal is strongly indicated.

Because of the many variables and unknowns associated with corrosion, a minimum allowance of $\frac{1}{8}$ in. is specified for carbon steel and low alloy steel even if no obvious corrosion or erosion problems exist. For higher alloys, such as stainless steel, a lower minimum of $\frac{1}{32}$ in. is often used.

The literature on corrosion is large and interpreting reported results is not always easy for the nonexpert. Advice from corrosion engineers should be sought, especially for unusual process environments differing in degree or kind from previous known cases. A useful compilation of data is given in Ref. 5. Reference to current literature and special reports and compilations of the National Association of Corrosion Engineers is recommended.

In the case of hydrogen service and services which cause pitting a corrosion allowance is not very useful. Hydrogen destroys metal strength by producing cracks or blisters, but the thickness of the metal is not reduced. At low temperature atomic hydrogen produced by thermal or catalytic dissociation diffuses into the metal along imperfections, ultimately recombining to form molecular hydrogen. The hydrogen pressure can increase to a point where it causes internal and surface blistering (6). At high temperatures hydrogen diffuses even more rapidly and forms methane by reacting with carbon content of the steels. The larger methane molecule builds up pressure that broduces high internal pressure and ultimately cracks (6,7). Neither of these processes reduces the metal thickness. Thus one selects for high temperature service a metal that will not be subject to attack, containing a carbide stabilizing element such as molydenum. Figure B.1 provides special data for hydrogen. The combination of H_2S and H_2 and H_2S alone require special care (see Ref. 8).

MATERIAL SELECTION

In addition to corrosion resistance a metal for pressure-vessel use should be selected to provide the highest allowable stress per unit cost. A cost study in 1970 (6) indicates that other factors being equal, carbon steel fulfills this need up to 825°F (440°C). Between 825°F and 1000°F (540°C) carbon $-\frac{1}{2}$% molydenum steel is attractive. When carbon steel is used up to 1000°F killed and semikilled flange and firebox grades are specified. Between 1000°F and 1200°F (540–650°C) various chromium-molydenum steels are closely grouped and issues other than strength are used to decide between them (5). Between 1200 and 1500°F (650–817°C) the austenitic stainless steels are the only metals for which stresses are given in the Code (2). Above 750°F one should also consider cold wall vessels or multilayer construction (3).

Table B.2 gives common steels used for reactors. Allowable stresses from the ASME Code for these frequently used steels are also listed.

MINIMUM THICKNESS AND OTHER FACTORS

Minimum thicknesses without allowance for corrosion as specified by the Code (2) for welded vessels are:

	Inches
Carbon and low alloy steel	$\frac{3}{32}$
Nonferrous metals (noncorrosive service) and high-alloy steels	$\frac{1}{16}$
Nonferrous metals (corrosive service) and high-alloy steels	$\frac{3}{32}$

Internal pressure usually governs the thickness of a reactor. Tall vessels should also be checked for wind loads (9).

EXAMPLE COST ESTIMATE

Estimate the cost of a reactor 10 ft ID × 26 ft operating at 26.53 atm and 794°F. Use ASTM A 387 Grade 22, Class 1 ($2\frac{1}{4}$ Cr-1 Mo) for resistance to hydrogen attack (see Fig. B.1).

Design pressure: $(26.53)(1.2)(14.7) = 467$ psig
Design temperature: 890°F (maximum catalyst use temperature)
$S = 13,100$ psi (Table B.2)
E, the joint efficiency = 1.0, for double butt welded and fully radiographed welds
Minimum corrosion allowance = $\frac{1}{8}$ in.

$$t = \frac{PR}{SE - 0.6P} = \frac{(467)(5)(12)}{(13,100)(1.0) - (0.6)(467)}$$

$$= 2.186 \text{ in.}$$

$$t_{actual} = 2.186 + 0.125 = 2.311 \text{ in}$$

Since this is an alloy steel, the calculated thickness will be used.

Table B.3 Properties of Vessel Heads

	2 : 1 Elliptical	Hemispherical	Standard ASME Torispherical
Capacity,[a] units $= L^3$	$\dfrac{\pi D^3}{24}$	$\dfrac{1}{12}\pi D^3$	$0.9\left[\dfrac{\pi D^2}{6}(\text{IDD})\right]$
IDD, inside depth of dish[a] (units $= L$)	$\dfrac{D}{4}$	$\dfrac{D}{2}$	$L-\left[(L-r)^2-\left(\dfrac{D}{2}-t-r\right)^2\right]^{\frac{1}{2}}$
Approximate weight[b] of dished portion of head (units $=$ mass)	$\rho\left[\dfrac{\pi(nD+t)^2 t}{4}\right]$	$\rho\,\dfrac{\pi D^2}{2}\,t$	$\rho\left[\dfrac{\pi\left(OD+\dfrac{OD}{24}+at\right)^2}{4}\right]$

Special Nomenclature: See Table B.1. Also ρ is the density of metal, $n = 1.20$ for $D = 60$ in. or less, 1.21 for $D = 62\text{–}76$ in., 1.22 for $D = 80\text{–}106$ in. and 1.23 for larger sizes, OD is the outside diameter, and $a = 2$ for thicknesses < 1 in. and 3 for thicknesses 1 in. and greater. D is the inside diameter.

[a] Does not include straight-flange portion on elliptical or torispherical heads. Include that volume in cylindrical portion of reactor.

[b] If weight of head including straight flanges is desired, add two times the straight flange length to the term inside the squared parenthesis. In estimating vessel weights, the straight flanged portion is included in cylindrical length.

Use 2 : 1 elliptical heads of same thickness

Vessel Weight: (ρ = 490 lb/cu ft)

$$\text{Shell, } (\pi)(10)\left(\frac{2.311}{12}\right)(26)(490) = 77079 \text{ lb}$$

Heads (see Table B.3)

$$(490)\left(\frac{\pi}{4}\right)\left[(1.23)(10) + \frac{2.311}{12}\right]^2\left(\frac{2.311}{12}\right)(2) = 23133 \text{ lb}$$

Total weight = 77,079 + 23,133 = 100,212 lb
For this size and type vessel a fabricated cost of 73¢/lb without nozzles
was suggested as an estimating figure (1971 cost) by a fabricator.

$$(100,212)(0.73) = \$73,155$$

Plus nozzles @ \$200/in. of diameter for this type vessel. The nozzles add
the following to the cost,

$$
\begin{array}{ll}
\text{2-20 in., } 2 \times 200 \times 20 = & 8000 \\
\text{1-12 in., } 1 \times 200 \times 12 = & 2400 \\
\text{4-1 \ in., } 4 \times 288 \qquad = & \underline{\ \ 800} \\
& \$11,200
\end{array}
$$

Often manufacturers will give an estimating cost per unit mass which
includes nozzles for a well-defined vessel.

REFERENCES

1. J. S. Clarke, *Hydrocarbon Process.*, **49** (6), 79 (1970).
2. *ASME Boiler and Pressure Vessel Code*, Section VIII, Division 1, American Society Mechanical Engineers, New York, 1974.
3. D. W. McDowell, Jr., J. O. Milligan, and A. D. Korin, *Hydrocarbon Process.*, **45** (3), 157 (1966).
4. G. Laithwaite, *Brit. Chem. Eng.*, **15** (7), 875 (1970).
5. G. A. Nelson, *Corrosion Data Survey*, Shell Development Co., 1960.
6. R. Q. Barr, *A Review of Factors Affecting the Section of Steels for Refining and Petrochemical Applications*, Climax Molydenum Co., Greenwich, Conn. (1971).
7. C. H. Samans, *Hydrocarbon Process.*, **42** (10), 169 and (11) 241 (1963).
8. S. L. Estefan, *Hydrocarbon Process.*, **49** (12), 85 (1970).
9. H. F. Rase and M. H. Barrow, *Project Engineering of Process Plants*, Wiley, New York, 1957.

APPENDIX C

Compressor Power and Costs

ISENTROPIC WORK

Enthalpy Method

$$(-W_s) = H_2 - H_1 = \Delta H_s \qquad \text{(C.1)}$$

where 1 and 2 indicate suction and discharge conditions, respectively. Equation C.1 may be solved conveniently by following a constant entropy line on an H–T–S diagram of the gaseous system in question. Such diagrams exist for various pure hydrocarbons, refrigerants, steam, natural gases, and H_2–N_2 mixtures. The actual discharge temperature is the temperature corresponding to the actual enthalpy $(H_2)_a$ at the discharge pressure.

$$(H_2)_a = \frac{\Delta H_s}{E_s} + H_1 \qquad \text{(C.2)}$$

where E_s is the isentropic efficiency and ΔH_s is the enthalpy change at constant entropy.

The actual work, W_a, is simply determined

$$W_a = \frac{\Delta H_s}{E_s}, \frac{\text{BTU or cal}}{\text{mole}}$$

hp = $(0.02358)(W_a, \text{BTU/lb mole})(\mathscr{F}_T, \text{lb moles/min})$, or 5.61×10^{-3} $(W_a, \text{cal/g mole})(\mathscr{F}_T, \text{g moles/sec})$.

Isentropic efficiencies for reciprocating compressors are given in Table C.1, but because of the wide deviation for centrifugal compression, polytropic efficiencies are obtained for each type of machine and corrected to isentropic

Table C.1 Typical Efficiencies for Reciprocating Compressors[a]

Compression Ratio	Engine-Driven[b]
1.1	50–60
1.2	60–70
1.3	65–80
1.5	70–85
2.0	75–88
2.5	80–89
3.0	82–90
4.0	85–90

Reproduced by permission: H. F. Rase and M. H. Barrow, *Project Engineering of Process Plants*, Wiley, New York, 1957.

[a] $bhp = \dfrac{\text{isentropic hp}}{\text{efficiency}}$

[b] Multiply the above values by 0.95 for motor-driven compressors.

(see the following) based on the gas and conditions in question using Fig. C.1. Typical polytropic efficiencies are given in Table C.2.

Ideal-Gas Method with Correction

For ideal gas

$$\int_1^2 v\,dP = \int_1^2 RT\,\frac{dP}{P} = \int_1^2 c_p\,dT$$

$$T_2 = T_1\left(\frac{P_2}{P_1}\right)^{R'/c_p} = T_1\left(\frac{P_2}{P_1}\right)^{(\kappa^\circ - 1)/\kappa^\circ} \tag{C.3}$$

where $\kappa^\circ = c_p/c_v$.

$$(-W_s) = \int_1^2 c_p\,dT = c_p(T_2 - T_1) = c_p T_1\left[\left(\frac{P_2}{P_1}\right)^{(\kappa^\circ - 1)/\kappa^\circ} - 1\right]$$

$$= \frac{\kappa^\circ}{\kappa^\circ - 1}RT_1\left[\left(\frac{P_2}{P_1}\right)^{(\kappa^\circ - 1)/\kappa^\circ} - 1\right], \frac{\text{energy units}}{\text{mole}} \tag{C.4}$$

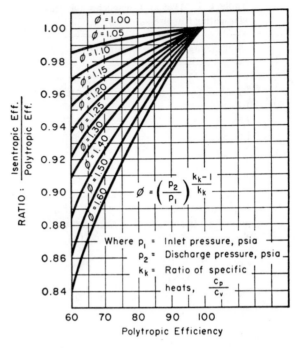

Fig. C.1 Relationship between isentropic and polytropic efficiency for centrifugal compressors. Adapted by permission, Clark Brothers Company and reproduced by permission: H. F. Rase and M. H. Barrow, *Project Engineering of Process Plants*, Wiley, New York, 1957.

Table C.2 Polytropic Efficiencies of Typical Process Compressors

Frame	*Normal Flow Range (icfm)	**Nominal Polytropic Head per Stage (Hp)	Nominal Polytropic Efficiency η_b	+Nominal Max. No. of Stages	Speed at Nominal Polytropic Head/Stage
29M	500- 8,000	10,000	.76	10	11,500
38M	6,000- 23,000	10,000/12,000	.77	9	8,100
46M	20,000- 35,000	10,000/12,000	.77	9	6,400
60M	30,000- 58,000	10,000/12,000	.77	8	5,000
70M	50,000- 85,000	10,000/12,000	.78	8	4,100
88M	75,000-130,000	10,000/12,000	.78	8	3,300
103M	110,000-160,000	10,000	.78	7	2,800
110M	140,000-190,000	10,000	.78	7	2,600
25MB (H) (HH)	500- 5,000	12,000	.76	12	11,500
32MB (H) (HH)	5,000- 10,000	12,000	.78	10	10,200
38MB (H)	8,000- 23,000	10,000/12,000	.78	9	8,100
46MB	20,000- 35,000	10,000/12,000	.78	9	6,400
60MB	30,000- 58,000	10,000/12,000	.78	8	5,000
70MB	50,000- 85,000	10,000/12,000	.78	8	4,100
88MB	75,000-130,000	10,000/12,000	.78	8	3,300

*Maximum flow capacity is reduced in direct proportion to speed reduction.
**Use either 10,000 ft or 12,000 ft for each impeller where this option is mentioned.
+At reduced speed, impellers can be added.

Courtesy Elliott Division Carrier Corp., Jeannette, Pa.

Units depend on value of R selected. For an actual gas the RT term becomes $z_m RT$, and Eq. C.4 must be multiplied by the average compressibility factor based on suction and discharge conditions.

For reciprocating compressors that do not deviate markedly from isentropic compression it is customary to base calculations on the isentropic discharge temperature as approximated by Eq. C-3 or determined from the H–T–S diagram.

The isentropic work for the actual gas is thus

$$(-W_s) = \frac{(z_m)_2 + (z_m)_1}{2} RT_1 \left[\left(\frac{P_2}{P_1} \right)^{(\kappa^\circ - 1)/\kappa^\circ} - 1 \right] \left(\frac{\kappa^\circ}{\kappa^\circ - 1} \right) \qquad (C.5)$$

In the case of a centrifugal compressor the temperature deviates markedly from the isentropic case. The compression is said to be polytropic and T_2 is obtained from Eq. C.3 or by the following relationship

$$T_2 = T_1 \left(\frac{P_2}{P_1} \right)^{(\kappa^\circ - 1)/\kappa^\circ E_p} \qquad (C.6)$$

where E_p is the polytropic efficiency.

The relationship between isentropic and polytropic efficiency becomes from Eq. C.5 and C.6

$$E_s = \frac{\Delta H_s}{\Delta H_a} \approx \frac{\left(\dfrac{P_2}{P_1} \right)^{(\kappa^\circ - 1)/\kappa^\circ} - 1}{\left(\dfrac{P_2}{P_1} \right)^{(\kappa^\circ - 1)/\kappa^\circ E_p} - 1} \qquad (C.7)$$

where ΔH_a is the actual enthalpy change. Equation C.7 is the basis for Fig. C.1. Thus for a centrifugal compressor the work is approximated by dividing Eq. C.5 by E_s or by using E_p in the denominator of the exponent of Eq. C.5.

NUMBER OF STAGES

Reciprocating

Compression ratios of 1.4 to 4.0 are common. When exceeded, use multiple staging as recommended by manufacturer with intercooling between stages. Theoretical optimum compression ratio is an equal value for each stage such that

$$\text{Compression Ratio per Stage} = \left(\frac{P_2}{P_1} \right)^{1/s} \qquad (C.8)$$

where s is the number of stages.

Centrifugal Compressors

Consult compressor manufacturer for maximum head per stage. As shown in Table C.2, a value of 10,000 ft-lb$_f$/lb$_m$ as polytropic head is common. The polytropic head is the actual head times the polytropic efficiency. If more than one machine is used, intercooling is specified. Temperature limits and energy economies may also indicate cooling between a group of stages within one frame (machine). Multiple machines are usually driven off a single shaft.

COMPRESSOR OPERATING COSTS

There is great merit in determining compression costs in terms of fuel cost since the price of fuel is usually an accurately known and economically sensitive variable. Electrically operated machines are of course conveniently considered in terms of local competitive power rates. In all cases estimates of driver power or fuel consumption should be obtained from the manufacturer. Gas turbine and gas engine fuel rates in the U.S.A. are usually given as BTU/bhp hr. Typical values are shown in Figs. C.2 and C.3. Electric motors operate in the range of 95 % efficiency in the usual large compressor application (see Table C.3).

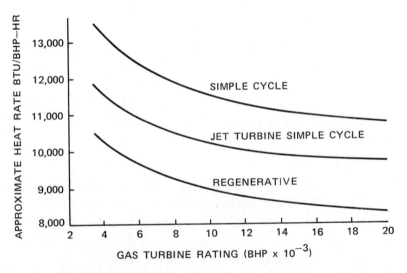

Fig. C.2 Approximate fuel rates for gas turbines. Reproduced by permission: S. B. Branch, *Hydrocarbon Process.*, **46** (10), 131 (1967).

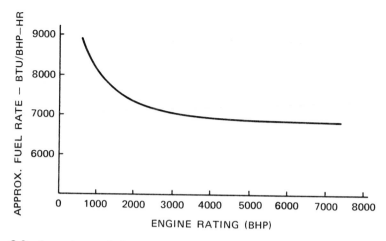

Fig. C.3 Approximates fuel rates for gas engines. Reproduced by permission: S. B. Branch, *Hydrocarbon Process.,* **46** (10), 131 (1967).

Table C.3   Full Load Efficiencies of Electric Motors

HP	3,600 rpm	1,200 rpm	600 rpm	300 rpm
5	80.0	82.5	—	—
	—	—	—	—
20	86.0	86.5	—	—
	—			82.7[a]
100	91.0	91.0	93.0	—
	—		91.4[a]	90.3[a]
250	91.5	92.0	91.0	—
	—	93.9[a]	93.4[a]	92.8[a]
1,000	94.2	93.7	93.5	92.3[a]
	—	95.5[a]	95.5[a]	95.5[a]
5,000	96.0	95.2	—	—
	—	—	97.2[a]	97.3[a]

Reproduced by permission: C. R. Olsen and E. S. McKelvy, *Hydrocarbon Process.,* **46** (10), 118 (1967).
[a] Synchronous Motors, 1.0 PF.

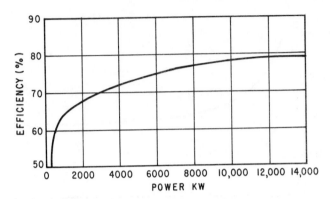

Fig. C.4 Approximate efficiencies of multistage turbines for process drives (multiply KW by 1.341 to convert to hp). Reproduced by permission: J. W. Farrow, *Hydrocarbon Process.*, **50** (3), 71 (1971).

Steam turbine consumption is expressed as a steam rate (SR) in, (mass of steam)/(hp hr). Values can be estimated from the theoretical isentropic steam rate corrected by turbine efficiency. Typical values are given in Fig. C.4.

$$SR = \frac{2545}{(H_1 - H_2)(E_t)}, \qquad \text{lb/hp hr}$$

where H_1 and H_2 are the inlet and discharge enthalpy of steam, BTU/lb and E_t is fractional turbine efficiency.

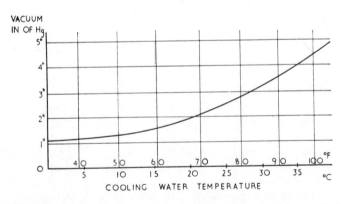

Fig. C.5 Practical vacuums for condensing turbines. Reproduced by permission: J. W. Farrow, *Hydrocarbon Process.*, **50**, 71 (1971).

Condensing turbines are specified for very large duties in order to minimize equipment size. The condensing pressure depends on the cooling water temperature (see Fig. C.5). The theoretical discharge enthalpy H_2 for a turbine corresponds to that of the vapor at the temperature and pressure of the discharge side, a value which must be found by following a constant entropy line on a Mollier diagram from inlet to outlet conditions. The actual enthalpy follows from the efficiency definition: $(H_2)_a = H_1 - (H_1 - H_2)(E_t)$. This enthalpy should be in the range of 8–12% moisture on the Mollier diagram. If this value is exceeded, other steam conditions must be selected. Excessive moisture causes rapid wear of the turbine blades.

The steam rate for a given turbine may be converted to equivalent fuel consumption in energy units per unit time as follows:

$$\text{Fuel consumption} = \frac{(SR)(\text{bhp})(H_1 - H_F)}{E_F}.$$

where H_F is the final enthalpy of the steam leaving system. In the case of a condensing turbine this would be the saturated water. In noncondensing systems it is the steam that enters the low pressure header and is H_2 less losses between turbine and header which are usually negligible. E_F is the generating facility fuel efficiency. Values in the range of 85% are typical.

APPENDIX D

Heat-Transfer Coefficients, Effective
Conductivities and Diffusitivities
in Packed Beds

HEAT-TRANSFER COEFFICIENTS AND EFFECTIVE CONDUCTIVITIES

The most precise data presented thus far is based on air flowing in beds with three types of catalysts (Table D.1)(1).

Wall Heat-Transfer Coefficient Based on Mean Temperature

$$\frac{q_r}{A_h} = h_i(T_m - T_w)$$

$$h_i = h_i^{\circ} + \frac{0.0005924}{D_p}\frac{D_p G}{\mu}, \quad 30 < \frac{D_p G}{\mu} < 1000 \tag{D.1}$$

where $h_i = \text{kcal/m}^2\text{hr}^{\circ}\text{C}$. This can be generalized

$$\frac{h_i D_p}{\lambda_g} = \frac{h_i^{\circ} D_p}{\lambda_g} + 0.024\frac{D_p G}{\mu_g} \tag{D.2}$$

where λ_g is the gas conductivity. Values for h_i° are given in Table D.2.

Table D.1 Properties of Catalyst Beds Used for Correlation

D (m)	D_p (m)	D/D_p	N_{Re}	Bed height (m)
Iron oxide catalyst for ammonia synthesis				
0.099	0.0095	10.4	200–900	0.3–1.0
0.1575	0.0095	16.6	90–400	0.4–1.4
Vanadium pentoxide catalyst for sulfuric acid production				
0.099	0.0059	16.7	60–600	0.4–1.0
0.1575	0.0059	26.7	25–300	0.4–0.8
Vanadium pentoxide catalyst for phthalic anhydride synthesis				
0.099	0.0057	17.4	150–400	0.3–1.0
0.1575	0.0057	27.6	50–260	0.2–1.1

Reproduced by permission: A. P. De Wasch and G. F. Froment, *Chem. Eng. Sci.*, **27**, 567 (1972).
All three catalysts are of cylindrical shape with height = dia.
In all cases the fluid is air.

Table D.2 Static Contributions to the Heat-Transfer Equations

	h_i°		λ_r°		h_w°	
	D_1	D_2	D_1	D_2	D_1	D_2
SO$_3$ cat	5.2	14.0	0.176	0.240	17.0	51.0
PA cat	6.6	13.3	0.241	0.224	16.2	70.0
NH$_3$ cat	10.0	20.5	0.360	0.350	31.0	85.0

Reproduced by permission: A. P. De Wasch and G. F. Froment, *Chem. Eng. Sci.*, **27**, 567 (1972).
$D_1 = 0.1575$ m.
$D_2 = 0.099$ m.

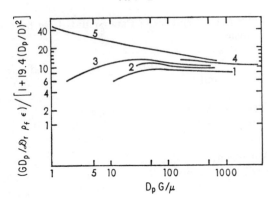

Fig. D.1   Correlations of radial effective diffusivity, $\mathscr{D}_r$ (numbers refer to data of different investigators). Reproduced by permission: G. F. Froment, *Ind. Eng. Chem.*, **59** (2), 18 (1967). [Copyright by American Chemical Society.]

Wall Heat-Transfer Coefficient Based on Temperature at $r = R_b$

$$\frac{q_r}{A_S} = h_w(T_{R_b} - T_w)$$

For air

$$h_w = h_w{}^\circ + 0.01152 \frac{D}{D_p}\left(\frac{D_p G}{\mu}\right) \tag{D.3}$$

Values for $h_w{}^\circ$ are given in Table D.2 where $h_w = \text{kcal}/(\text{m}^2)(\text{hr})(^\circ\text{C})$.

Effective Thermal Conductivity, λ_r

For air

$$\lambda_r = \lambda_r{}^\circ + \frac{0.0025}{1 + 46\left(\dfrac{D_p}{D}\right)^2} \frac{D_p G}{\mu} \tag{D.4}$$

Values for $\lambda_r{}^\circ$ are given in Table D.2 where λ_r is the effective conductivity in kcal/(m)(hr)(°C).

Relationship between h_i, h_w, and λ_r (see p. 541¹)

$$h_i = \frac{1}{\dfrac{1}{h_w} + \dfrac{D}{8\lambda_r}} = \frac{h_w}{1 + \dfrac{h_w D}{8\lambda_r}} \tag{D.5}$$

Relationship between Static Contributions

$$h_i{}^\circ = \frac{2.583\lambda_r{}^\circ}{D^{1.33}} \tag{D.6}$$

$$h_w{}^\circ = \frac{8.4\lambda_r{}^\circ}{D^{1.33}} \tag{D.7}$$

where D is in meters.

Since values of $\lambda_r{}^\circ$ are not so dependent on tube diameter (see Table D.2), it seems preferable to select a value of $\lambda_r{}^\circ$ and obtain $h_w{}^\circ$ or $h_i{}^\circ$ from Eqs. D.6 and D.7.

With reference to previous studies (2–4) of packed-bed heat transfer, the following equations, based on Eqs. D.2–D.4, can be argued as reasonable for all gases. Data on other gases, however, are rare.

$$\frac{h_i D_p}{\lambda_g} = \frac{h_i{}^\circ D_p}{\lambda_g} + 0.0324 \left(\frac{c_p \mu}{\lambda_g}\right)\left(\frac{D_p G}{\mu_g}\right) \tag{D.8}$$

$$\frac{h_w D_p}{\lambda_g} = \frac{h_w{}^\circ D_p}{\lambda_g} + 0.631D \left(\frac{c_p \mu}{\lambda_g}\right)\left(\frac{D_p G}{\mu_g}\right) \tag{D.9}$$

Equation D.9 is a dimensional equation in D with units of meters.

$$\frac{\lambda_r}{\lambda_g} = \frac{\lambda_r{}^\circ}{\lambda_g} + \frac{0.137}{1 + 46\left(\dfrac{D_p}{D}\right)^2} \left(\frac{c_p \mu}{\lambda_g}\right)\left(\frac{D_p G}{\mu_g}\right) \tag{D.10}$$

RADIAL EFFECTIVE DIFFUSIVITY OF PACKED BEDS

The data are not very precise, as shown in Fig. D.1, but models are not very sensitive to values of this parameter.

REFERENCES

1. A. P. de Wasch and G. F. Froment, *Chem. Eng. Sci.*, **27**, 567 (1972).
2. S. Yagi and D. Kunii, *A.I.Ch.E.J.*, **3**, 373 (1957).
3. S. Yagi and D. Kunii, *A.I.Ch.E.J.*, **6**, 97 (1960).
4. D. G. Bunnell, H. B. Irwin, R. W. Olson, and J. M. Smith, *Ind. Eng. Chem.*, **41**, 1977 (1949).

APPENDIX E

Selected Sources of Data

1. *Thermodynamic Properties*

    (a) D. R. Stull, E. F. Westrum, Jr., and G. L. Sinke, *The Chemical Thermodynamics of Organic Compounds*, Wiley, New York, 1969.
    (b) *Selected Values of Properties of Hydrocarbons and Related Compounds*, API Research Project 44 and *Selected Values of Properties of Chemical Compounds*, TPRC Project, Consists of loose-leaf data sheet compilation on a continuing basis, Thermodynamics Research Center, Texas A & M, College Station, Tex.
    (c) D. R. Stull and H. Prophet, *JANAF Thermochemical Tables*, U.S. Government Printing Office, Washington, D.C., 1971.
    (d) API, *Technical Data Book–Petroleum Refining*, American Petroleum Institute, Washington, D.C., 1966.

2. *Diffusivities (Gaseous)*

    (a) R. C. Reid and T. K. Sherwood, *The Properties of Gases and Liquids: Their Estimation and Correlation*, 2nd ed., McGraw-Hill, New York, 1966.
    (b) T. R. Marrero and E. A. Mason, *J. Phys. Chem. Ref. Data*, **1**, 3 (1972). Summary of references through 1970.
    (c) S. Bretsznajder, *Prediction of Transport and other Physical Properties*, Pergamon, New York, 1971.
    Also Ref. 1d.

3. *Diffusivities (Liquid)*

    (a) D. M. Himmelblau, *Chem. Rev.*, **64**, 527 (1964).
    See also 1d, 2a, and 2c.

4. *Viscosity*

   (a) *Standards Tubular Exchanger Manufacturers Association*, T.E.M.A., New York (latest edition)
   Also Ref. 1d (good method for mixtures) and 2a.

5. *Thermal Conductivity*

   (a) Y. S., Touloukian, et al., *Thermophysical Properties of Matter* (10 volumes), Plenum, New York, 1970.
   (b) R. A. Sehr, *Chem. Eng. Sci.*, **9**, 145 (1958). (for typical catalysts).
   Also Ref. 1d, 2a, and 4a

6. *Effective Catalyst Conductivity and Diffusivity*

   (a) C. N. Scatterfield, *Mass Transfer in Heterogeneous Catalysis*, MIT Press, Cambridge, Mass., 1970.

7. *Vapor Pressure*

   (a) B. J. Zwolinski and R. C. Wilhoit, *Handbook of Vapor Pressures and Heats of Vaporization of Hydrocarbons and Related Compounds*, Thermodynamics Research Center, College Station, Tex. 1971.
   (b) T. E. Jordan, *Vapor Pressure of Organic Compounds*, Interscience, New York, 1954.

8. *General Physical Properties Estimation*

   (a) R. Gallant, *Physical Properties of Hydrocarbons*, Vols. 1 and 2, Gulf, Houston, 1968.
   See also 2(a) for estimating methods.

APPENDIX F

Tube-Side Heat Transfer Coefficient for Water

Table C.2

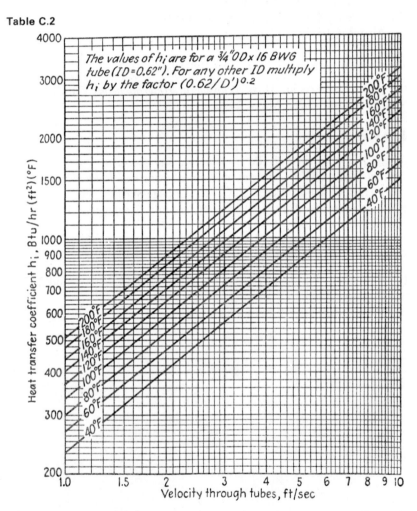

The values of h_i are for a $\frac{3}{4}"OD \times 16\ BWG$ tube (ID=0.62"). For any other ID multiply h_i by the factor $(0.62/D')^{0.2}$

Heat transfer coefficient h_i, Btu/hr (ft²)(°F)

Velocity through tubes, ft/sec

APPENDIX G

Inert Catalyst-Bed Supports

	Type	
	Ceramic[a]	Alumina
Shapes	balls or cylindrical pellets	balls or crushed lumps
Bulk density, lb/ft^3	87–92	119–130
Maximum use temp., °F	1500–2000	3450
Specific gravity	2.4	3.95
Composition	38.1% Al$_2$O$_3$ 54.4% SiO$_2$	94.5%[b] Al$_2$O$_3$
Hardness (Mohs scales)	6.5	9.0
Typical sizes (ball), in.	$\frac{1}{4}, \frac{3}{8}, \frac{1}{2}, \frac{5}{8}, \frac{3}{4}, 1,$ $1\frac{1}{4}, 1\frac{1}{2}, 2$	$\frac{1}{4}, \frac{3}{8}, \frac{1}{2}, \frac{3}{4}$
(Pellets), in.	$\frac{1}{8} \times \frac{1}{8}, \frac{1}{4} \times \frac{3}{8},$ $\frac{3}{8} \times \frac{1}{2}, \frac{5}{8} \times \frac{3}{4},$ $\frac{3}{4} \times \frac{7}{8}, \frac{7}{4} \times 1\frac{1}{4}$	
(Lumps), in.		$\frac{1}{8}, \frac{3}{8}, \frac{3}{8}$ to $\frac{5}{8},$ $\frac{5}{8}$ to $\frac{7}{8}, \frac{3}{4}$ to $1\frac{1}{4},$ $1\frac{1}{4}$ to $1\frac{3}{4}$
Relative price factor	1	3

[a] Based on product of Norton Chemical Process Division.
[b] Grades up to 99.3% Al$_2$O$_3$ with low silica and iron and use temperatures up to 3600°F available.

NOMENCLATURE

Units are described by the following symbols:

H = thermal energy t = time
L = length T = temperature
M = mass

A	frequency factor, units same as corresponding rate constant
A, B, ...	reactants
A_c	cross-sectional area, L^2
A_D	acceptable percentage deviation of rate from that at wall
A_f	absorption factor, dimensionless, see Eq. 14.55
A_h	heat-transfer surface, L^2
a, b, c, ..., r, s, ...	stoichiometric coefficients
$a, b, c, ..., r, s, ...$	exponents in rate equations and various constants
$\mathbf{a}_A, \mathbf{a}_B, \mathbf{a}_C, ...$	activities of indicated components
a_c	external surface area of catalyst per unit volume of slurry, L^{-1}
a_m	external surface area of catalyst per unit mass, $L^2 M^{-1}$
a_p	external surface area of an average particle, L^2
a_s	geometric surface area per unit volume of particle, L^{-1}
a_t	total surface area of packing per unit volume of packed system, L^{-1}
a_V, a_v	surface area of dispersed phase (bubble or drop) per unit volume of system or expanded slurry and per unit volume of continuous phase, respectively, L^{-1}
a_w	wetted area of catalyst particle per unit mass of catalyst, $L^2 M^{-1}$
a_{WR}	surface area per unit volume of screen wire, L^{-1}
B	$E/R'T_j^2$ or $E/R'T_w^2$
b_{ch}	width of channel between plates, L
b_{ck}	thickness of coke, L
b_I	mass of inert packing per mass of catalyst
b_s	screen thickness, L
b_v	volume of inert packing per unit volume of catalyst
b_w	wall thickness, L

C	concentration, $(\text{moles})L^{-3}$
$C^{\ddagger}$	concentration of activated complex, $(\text{moles})L^{-3}$
$C_A, C_B, \ldots, C_R, C_S$	molar concentration of indicated component, $(\text{moles})L^{-3}$
$C_A{}^*$	concentration of A in bulk liquid at chemical equilibrium, $(\text{moles})L^{-3}$
$C_A{}', C_B{}', \ldots, C_R{}', C_S{}'$	moles of indicated component adsorbed per unit mass, $(\text{moles})M^{-1}$
$C_A{}''$	concentration of A at beginning of poisoned zone, $(\text{moles})L^{-3}$
$C_{A_a}, C_{B_b}, \ldots$	molar concentration of indicated component in indicated phase a or b, $(\text{moles})L^{-3}$
C_{B_b}	molar concentration of B in bulk liquid, $(\text{moles})L^{-3}$
C_{B_d}	molar concentration of B in dispersed phase, $(\text{moles})L^{-3}$
C_{B_c}	molar concentration of B in continuous phase, $(\text{moles})L^{-3}$
$C_{A_e}, C_{B_e}, \ldots, C_{R_e}, C_{S_e}$	effluent concentration of indicated component, $(\text{moles})L^{-3}$
$C_{A_F}, C_{B_F}, \ldots$	concentration of indicated component in feed, $(\text{moles})L^{-3}$
$C_{A_I}, \ldots, C_{j_I}$	concentration of indicated component inside catalyst particle, $(\text{moles})L^{-3}$
$C_{A_i}, C_{B_i}, \ldots, C_{R_i}, C_{S_i}$	molar concentration of indicated component at interface between two fluid phases, $(\text{moles})L^{-3}$
$C_{A_s}, C_{B_s}, \ldots$	concentration of indicated component at outer surface of catalyst, $(\text{moles})L^{-3}$
C_D, C_{ds}, C_{dst}	drag constants (see Eqs. 13.19–13.23)
$C_l{}'$	concentration of vacant sites per unit mass, $(\text{moles})M^{-1}$
c_d	loss coefficient for orifice, screen, or perforated plate
c_o	hopper cone angle correction (see Fig. 13.9)
c_p	heat capacity at constant pressure for mixture in mass units, $HM^{-1}T^{-1}$ (f superscript designates fluid when possibility for ambiguity exists)
$c_{pA}, c_{pB}, \ldots, c_{pR}, c_{pS}$	molal heat capacity at constant pressure for indicated component, $H(\text{moles})^{-1}T^{-1}$
c_{p_a}	molal heat capacity at constant pressure of reaction mixture per mole of A fed, $H(\text{mole of A})^{-1}T^{-1}$
c_{pF}	heat capacity of feed, $HM^{-1}T^{-1}$

c_{pr}	heat capacity of catalyst or other solid particle, $HM^{-1}T^{-1}$
c_{pm}	molal heat capacity at constant pressure of mixture, $H(\text{moles})^{-1}T^{-1}$
c_v	molal heat capacity at constant volume, $H(\text{moles})^{-1}T^{-1}$
$\mathbf{c_w}$	wall effect correction (see Eq. 13.15)
D	reactor or pipe inside diameter, L
D'	inside diameter in inches of reactor tube
$\overline{D}$	mean diameter of tube
$\mathbf{D_{AB}}, \mathbf{D_{BC}}, \dots$	bond energy for indicated bonds, $H(\text{mole})^{-1}$
D_c	diameter of coil, L
D_{ch}	diameter of channel, L
D_{ck}	inside diameter of cylinder of coke, L
$\overline{D}_{ck}$	mean diameter of coke cylinder, L
D_I	impeller diameter, L
D_H, D_H'	equivalent diameters for reactor jackets for heat transfer and pressure drop, respectively, L
D_j	diameter of jet, L
D_m	mean diameter of particles from screen or microscopic analysis, L
D_O	outside diameter of pipe or tube, L
$\overline{DP}$	degree of polymerization
D_p	diameter of a sphere having the same a_s as the particle in question, L
D_s	shaft diameter, L
$\mathscr{D}$	molecular diffusivity, $L^2 t^{-1}$
$\mathscr{D}_A, \mathscr{D}_B, \dots$	molecular diffusivity of indicated component, $L^2 t^{-1}$
$\mathscr{D}_a$	axial dispersion coefficient (axial diffusivity), $L^2 t^{-1}$
$\mathscr{D}_f$	dispersion coefficient in dense phase of two-phase fluidized-bed model, $L^2 t^{-1}$
$\mathscr{D}_{I_A}, \mathscr{D}_{I_B}, \dots$	effective diffusivity in the interior of catalyst particle for indicated component, $L^2 t^{-1}$
$\mathscr{D}_r$	effective radial diffusivity in catalyst bed, $L^2 t^{-1}$
d	diameter, L (d_1 = diameter of pipe jacket)
d_B	bubble diameter, L
$\mathbf{d_c}$	diameter of cylinder, L
d_{ct}	outside diameter of coil tube, L
d_d	average drop diameter, L
d_{max}	maximum drop diameter, L
d_{min}	minimum drop diameter, L

d_p	diameter of a sphere having same surface area as particle in question, L
d_s	screen pore diameter, L
d_{sp}	diameter of sphere
d_v	diameter of sphere having same volume as particle in question, L
d_w	diameter of warp wire, L
E	energy of activation, $H(\text{mole})^{-1}$; prime indicates value for reverse reaction; subscript *obs* indicates observed value when intraparticle diffusional resistances are significant
E_B	fraction of bed that is effective
E_c	contactor efficiency
E_D	diffusional activation energy, $H(\text{mole})^{-1}$
E_F	fuel efficiency for generating facility
E_G	fractional point efficiency in plate tower
E_{mv}	fractional Murphee plate efficiency
E_p	polytropic efficiency
E_s	isentropic efficiency for compressor
E_T	energy dissipation per unit mass
E_t	turbine efficiency
E_z	energy necessary to separate two drops; see Eq. 15.13
EA	equilibrium approach, fractional approach to equilibrium
$e, f, g, i, \ldots$	exponents in rate equations not necessarily the same as stoichiometric coefficients
$F_A, F_B, \ldots$	molar feed rate of indicated component, $(\text{moles})t^{-1}$
F_e	outlet flow from reactor, moles t^{-1}
F_g	molar gaseous feed rate, $(\text{moles})t^{-1}$
F_L	molar liquid feed rate, $(\text{moles})t^{-1}$
F_{PU}	molar purge rate, $(\text{moles})t^{-1}$
F_Q	molar quench flow rate, $(\text{moles})t^{-1}$
F_r	molar recycle flow rate, $(\text{moles})t^{-1}$
F_T	total molar feed rate, $(\text{moles})t^{-1}$
$F(t)$	residence time distribution function (see p. 318[1])
$\mathscr{F}_A, \mathscr{F}_B, \mathscr{F}_j, \ldots, \mathscr{F}_T$	moles of indicated component flowing per unit time at any position Z, $(\text{moles})t^{-1}$; T refers to total moles
f	friction factor for open pipe, dimensionless
$f_A, f_B, \ldots$	fugacity of indicated component, atm
$f_A^\circ, f_B^\circ, \ldots, f_j^\circ$	standard-state fugacity for indicated component (unity for gaseous systems), atm

f_e	initiator efficiency
f_g	friction factor for gas in fluidized bed or in pipe
f_k	friction factor for packed bed
f_p	particle friction factor in fluidized bed
f_w	wire area per gauze cross-sectional area
G	mass velocity, $Mt^{-1}L^{-2}$
G'	molar mass velocity, $(moles)t^{-1}L^{-2}$
ΔG°	standard free energy of reaction, $H(mole)^{-1}$
$\Delta G^\ddagger$	free energy of activation, $H(mole)^{-1}$; subscript L refers to standard state of pressure and temperature of system; subscript p refers to standard state at 1 atm
G_g	gaseous mass velocity, $Mt^{-1}L^{-2}$
G_I	insoluble or nonabsorbable (inert) gas mass velocity, $Mt^{-1}L^{-2}$
G_L	liquid mass velocity, $Mt^{-1}L^{-2}$
G_{mf}	mass velocity at minimum fluidization velocity, $Mt^{-1}L^{-2}$
G_p	solids superficial mass velocity, $Mt^{-1}L^{-2}$ (subscript ch designates velocity at choking conditions)
G_s	mass velocity in $lb/(ft)^2$ (sec)
g	acceleration due to gravity
g_c	force–mass conversion factor, 32.17 $(lb_m ft)/(lb_f sec^2)$ and 980.7 $(g_m cm)/(g_f sec^2)$
g_f, g_m	grams of force and grams of mass, respectively
H	actual enthalpy at pressure and temperature of systems, $H(mole)^{-1}$, subscript $(H_A, H_B, H_j, \ldots)$ indicates component, or position
H^*	ideal gas enthalpy, $H(mole)^{-1}$
ΔH°	standard heat of reaction, $H(moles)^{-1}$
$\Delta H^\ddagger$	enthalpy of activation, $H(mole)^{-1}$
H_a	holdup for phase **a** or acid phase, volumes of **a** per unit volume of total mixture
$(\Delta H_A)_{T_1}$	heat of reaction at indicated temperature, per mole of A reacted, $H(mole)^{-1}$
ΔH_a	heat of adsorption, $H(mole)^{-1}$; primed value is differential heat of adsorption; *iso* indicates isoteric
H_b	holdup for **b** phase, volumes of **b** per total volume
H_D	head generated by impeller, ft lb_f/lb_m or cm g_f/g_m
H_d	fractional volume or holdup of dispersed phase, volume of dispersed phase per unit volume of total mixture

$H_{f_A}{}^\circ, H_{f_B}{}^\circ, \ldots, H_{f_j}{}^\circ$	standard heat of formation per mole of indicated component at 298°K, $H(\text{mole})^{-1}$
H_g	fractional gas holdup, volume of gas per unit volume of operating reactor volume
H_L	total liquid holdup fraction, volume of liquid per total volume
H_l	enthalpy loss per mass of oil flow, HM^{-1}
H_{OL}	operating fractional liquid holdup, volume of liquid per unit volume of operating reactor volume
ΔH_s	isentropic enthalpy change
ΔH_v	latent heat of vaporization
h	heat-transfer coefficient, $Ht^{-1}L^{-2}T^{-1}$
$\mathbf{h}$	Planck's constant, 6.624×10^{-27} erg-sec per molecule
h_c	height of cylinder, L
h_G	gas heat-transfer coefficient, $Ht^{-1}L^{-2}T^{-1}$
h_g	liquid heat-transfer coefficient with sparging, $Ht^{-1}L^{-2}T^{-1}$
h_i	inside film coefficient of tube, $Ht^{-1}L^{-2}T^{-1}$
h_L	liquid heat-transfer coefficient, $Ht^{-1}L^{-2}T^{-1}$
h_l	liquid height on tray, L
h_o	outside film heat-transfer coefficient
h_s	heat-transfer coefficient between catalyst surface and surrounding fluid, $Ht^{-1}L^{-2}T^{-1}$
h_T	overall heat-transfer coefficient from position $\mathbf{r} = R_b$ of bed to the jacket fluid, $Ht^{-1}L^{-2}T^{-1}$
h_w	wall heat-transfer coefficient, see page 406[1] Fq. 9.16, $Ht^{-1}L^{-2}T^{-1}$
$h_w{}^s, h_w{}^f$	wall heat-transfer coefficient for solid and fluid, respectively, $Ht^{-1}L^{-2}T^{-1}$
hp	horsepower
$I, \mathbf{I}$	intercept and inert, respectively
$[I]_a$	carbonium ion concentration in acid phase
I_m	dispersed phase mixing modulus
J	mechanical equivalent of thermal energy, 778 ft-lb$_f$/BTU, 0.4267 kg$_f$ m/cal; $1/J = 1.286 \times 10^{-3}$ BTU/ft-lb$_f$, 2.343 cal/kg$_f$ m
J_D	J-factor for mass transport at catalyst surface, dimensionless (see Eq. 11.15)
J_h	J-factor for heat transfer at catalyst surface, dimensionless (see Eq. 11.16)
$\mathbf{J_R}$	fraction of collisions that lead to reaction

J_w	jacket space, L (see Table 8.7)
j	any component
K	thermodynamic reaction equilibrium constant
K'	equilibrium constant for the surface reaction
$K^{\ddagger}$	thermodynamic equilibrium constant between reactants and activated complex in terms of activities, dimensionless
$K_A, K_B, K_R, K_S, \ldots$	apparent adsorption equilibrium constants for the indicated reactants, atm^{-1} for gaseous components or $(moles)^{-1}L^3$ when concentration is used in rate equation
K_b	concentration equilibrium constant for reaction in phase "b" (units depend on stoichiometry)
K_c	equilibrium constant in concentration units (units depend on stoichiometry)
$K_c^{\ddagger}$	concentration equilibrium constant between reactants and complex that does not include the partition function for the reaction coordinate
K_D, K_D°	dissociation constants for substituted and unsubstituted aromatic, respectively
$(K_d)_A, (K_d)_B$	distribution coefficients for indicated components, dimensionless
K_G	overall mass-transfer coefficients from bulk gas to bulk liquid phase in terms of partial-pressure driving force, $(moles)t^{-1}L^{-2}(atm)^{-1}$
K_L	overall mass-transfer coefficients from bulk gas to bulk liquid phase in terms of concentration driving force, Lt^{-1}
$\mathbf{K}_n$	controller gain
K_p	equilibrium constant in partial-pressure units (units depend on stoichiometry)
$\mathbf{K}_{th}$	diffuser throw constant (see Table 7.3)
K_v	vapor–liquid equilibrium constant, dimensionless
k	rate constant as designated at point of use or based on activities
$\mathbf{k}$	Boltzmann constant, 1.3805×10^{-16} erg/(°K) (molecule)
$k_A, k_B, \ldots, k_R, k_S, \ldots$	adsorption rate constants for indicated components (desorption is indicated by prime), $(moles)L^{-3}t^{-1}(atm)^{-1}$ $[(moles)L^{-3}t^{-1}$ for $k_A']$
k_a, k_b	reaction rate constants for multiphase reactions in a and b phases, respectively

$k_{aa}, k_{ab}, k_{ba}, k_{bb}$	copolymerization rate constants (see Eqs. 4.20 and 4.21)
k_c	reaction rate constant for rate equations involving concentration terms (prime indicates reverse rate constant), $(L^3/\text{moles})^{n-1}t^{-1}$
$\hat{k}_c$	same as k_c except per unit mass of catalyst, $L^{3n}(\text{moles})^{1-n}M^{-1}t^{-1}$
k_{c_e}	same as k_c except based on exterior area of catalyst, $(\text{moles})^{1-n}L^{3n-2}t^{-1}$
k_{c_s}	reaction rate constant in terms of total catalyst surface area, $(\text{moles})^{1-n}L^{3n-2}t^{-1}$
k_{c_v}	reaction rate constant based on volume of catalyst and concentrations, $(\text{moles})^{1-n}L^{3n-3}t^{-1}$
k_d	rate constant for initiator dissociation
$k_{d0}, k_{d1}, \ldots, k_{dm}$	deactivation rate constants, order indicated by second subscript, that is, k_{d1} = first-order deactivation constant
k_f	a concentration independent reaction rate constant for a gaseous reaction, units based on order
k_G	mass-transfer coefficient between bulk gas and gas–liquid interface, partial-pressure driving force, $(\text{moles})t^{-1}L^{-2}(\text{atm})^{-1}$
$k_{g_A}{}^s, k_{g_B}{}^s$	mass-transfer coefficient for indicated component between catalyst surface and surrounding fluid with pressure driving force, $(\text{moles})t^{-1}L^{-2}(\text{atm})^{-1}$
k_L	mass-transfer coefficient between bulk liquid and gas or liquid–liquid interface when reaction occurs, Lt^{-1}
$k_L{}^\circ$	mass-transfer coefficient between bulk liquid and liquid–gas interface without reaction occurring, Lt^{-1}
k_n	rate constant for nth order reaction
k_O	apparent overall rate constant (see Eqs. 14.23 and 14.24)
k_p	reaction rate constant in terms of partial-pressure units; prime indicates reverse reaction, $(\text{moles})L^{-3}t^{-1}(\text{atm})^{-n}$; $k_p{}^\bullet$ is value of k_p at outlet temperature
$\hat{k}_p$	reaction rate constant in terms of catalyst mass and partial pressures, $(\text{moles})M^{-1}t^{-1}(\text{atm})^{-n}$
k_{pr}	polymerization (propagation) rate constant
k_{p_s}	reaction rate constant based on total surface area and partial-pressure units, $(\text{moles})L^{-2}t^{-1}(\text{atm})^{-n}$

k_{p_e}	reaction rate constant based on exterior surface area of particle and in terms of partial pressures, (moles)(exterior surface area)$^{-1}t^{-1}(atm)^{-n}$
k_q	concentration independent rate constant for systems forming non-ideal solutions
k_s	surface reaction rate constant (prime is used for reverse reaction), units depend on order
k_{s_A}, k_{s_B}	mass-transfer coefficient between catalyst surface and surrounding fluid with concentration driving force for indicated component, Lt^{-1}
k_t	reaction rate constant for termination
k_{tc}	rate constant for termination by combination
k_{tr}	rate constant for chain transfer
L	reactor length, or length of bed in a catalytic reactor or packed tower, L
LHSV	liquid hourly space velocity, (volumes liquid)(volume reactor)$^{-1}t^{-1}$
$L_A, L_B, L_C, \ldots$	flammability limit in percent of indicated component
L_B	equivalent length of one return bend
L_d	length of diffuser wall, L
L_e	equivalent length of tube, ft
L_K	thickness of flat plate, L
L_n	length of straight pipe, L
L_p	depth of pore, L
L_R	concentration of active sites, (moles)M^{-1}
l_1	impeller blade length
lb_f, lb_m	pounds of force and pounds of mass, respectively
ℓ	symbol for an active site
l_{bp}	distance between baffles in jacket. L
l_s	scale of smallest eddies, L
M	monomer
$M_A, M_B, M_j, \ldots$	molecular weight of indicated component
M_F	molecular weight of total feed
M_g	molecular weight of gas phase
M_I	molecular weight of inerts
M_m	molecular weight of mixture
$\overline{M}_n$	number average molecular weight
m	variously used as slope, constant, exponent, and deactivating event order
m_c	catalyst loading in slurry reactor per unit volume of expanded slurry, ML^{-3}
m_d	mass of monomer in drops, m

m_E	allowable entrainment, Mt^{-1}
m_H	Henry's law constant
m_o	oil flow rate, Mt^{-1}
m_s	mass of sample, M
m_{sf}	solids flow rate, Mt^{-1}
m_Z	rate of solids thrown into freeboard, Mt^{-1}
N	impeller revolutions per unit time, usually expressed as rpm
$N_A, N_B, \ldots$	mass-transfer rate of indicated component, (moles) $L^{-2}t^{-1}$
$N_A', N_B', \ldots$	mass-transfer rate, (moles) t^{-1}
N_{BO}	$(k_g)_p R_p / \mathscr{D}_{I_p}$, Biot number for poison precursor
N_{DA}	$(k_c)_p R_p / \mathscr{D}_{I_p}$, Damkohler number for poison
N_{Fr}	impeller Froude number, $N_2 D_1^2 / g_c$
N_{Kr}	Karlovitz number (see Eq. 10.32)
N_{Nu}	Nusselt number defined as used
N_o	Avogadro's number, 6.02252×10^{23} molecules/mole
N_p	power number, $P_a g_c / \rho N^3 D_1^5$
$(N_{Pe})_a$	axial Peclet number, $D_p u_s / \mathscr{D}_a$ for beds and $Du / \mathscr{D}_a$ for empty tubes
$(N_{Pe})_r$	$D_p u_s / \mathscr{D}_r$, radial Peclet number
N_{Pr}	Prandtl number, $c_p \mu / \lambda_f$
N_{Q_R}	discharge coefficient, Q_R / ND_1^3
N_{Re}	Reynolds number, $D_p G / \mu$ for catalyst bed, DG / μ for empty tube; $\bar{d}_v u_{mf} \rho_g / \mu_g$ for fluidized bed at minimum fludization velocity
$(N_{Re})_I$	impeller Reynolds number, $\rho ND_1^2 / \mu$
N_s	solid diffusion number, for poison $3\mathscr{D}_{I_p} L / R_p^2 u_i$ where $\mathscr{D}_{I_p}$ is the effective diffusivity in particle of poison precursor
N_{Sc}	Schmidt number, $\rho \mu_f / \mathscr{D}_j$
N_{St}	Stanton number, $h / c_p G$
N_T	total number of trays
$N_\alpha, N_\beta, N_\delta, N_\zeta$	dimensionless groups for fluidized bed (see Eqs. 13.30–13.33)
n	reaction order, also used to designate number of carbon atoms in a molecular formula
$n_A, n_B, \ldots, n_I$	moles of indicated component (n_I refers to inerts)
$n_{A_a}, n_{B_a}, \ldots, n_{I_a}$	moles of indicated component per mole of A fed (I refers to inerts)
$n_{A_F}, n_{B_F}, \ldots$	moles of indicated component per total moles of total feed
n_b	number of impeller blades

n_{bv}	number of vertical baffles
n_{ch}	number of channels per unit cross section of bed, L^{-2}
n_I	number of equally spaced impellers
n_m	number of monomer units
n_r	number of reactors in series
n_s	number of gauzes or screens
n_{sg}	number of slugs per unit volume of fluidized bed
n_T	total moles in reaction mix or total moles per mole of A, n_{T_a}, or per total moles of feed, n_{T_F}
n_t^*	total moles per unit mass at equilibrium, (moles) M^{-1}
n_{tr}	moles of tracer added in pulse
n_W	mesh size of screen, (wires)L^{-1}
P	total pressure, atm or (force)L^{-2}, similar units for other pressure terms unless stated otherwise
ΔP	pressure drop
$\Delta P'$	pressure drop, *psi*
$P_A, P_B, P_j, \ldots, P_R, P_S$	partial pressure for indicated component, atm; asterisk is used to indicate value at thermodynamic equilibrium; sub I indicates inerts
P_{As}, P_{Bs}	partial pressure of indicated component at catalyst surface, atm
P_a	agitator power, (force)Lt^{-1}
P_{cr}	critical pressure
ΔP_d	pressure gradient due to drag
ΔP_D	distributor pressure drop
P_e	equivalent power (see Table 14.4)
ΔP_e	pressure drop due to sudden enlargement
P_f	pressure factor, $y_f P$
$(\Delta P_f)_p, (\Delta P_f)_g$	frictional pressure drop due to particles and fluid, respectively, in a fluidized bed
$\Delta P_G, (\Delta P/\Delta Z)_G$	pressure drop and gradient as if gas flowing alone
P_g	power dissipated in liquid by sparging, (force)Lt^{-1}
P_k	power required to compress gas; see Eq. 14.10
$\Delta P_L, (\Delta P/\Delta Z)_L$	pressure drop and gradient as if liquid flowing alone
P_o	vapor pressure, atm
$\Delta P_{TP}, (\Delta P/\Delta Z)_{TP}$	two-phase flow pressure drop and gradient
$\mathbf{p}$	steric factor in collision theory
p	impeller blade pitch
Q	volumetric flow rate, $L^3 t^{-1}$
Q_F	volumetric feed rate, $L^3 t^{-1}$

Q_g	volumetric gas flow rate, L^3t^{-1}
Q_L	volumetric liquid flow rate, L^3t^{-1}
Q_R	volumetric circulation rate in stirred tank or recycle rate, L^3t^{-1}
q	heat-transfer rate, Ht^{-1}
q_{diff}	differential heat of adsorption, $H(mole)^{-1}$
q_g	heat generated, Ht^{-1} (prime indicates per mole of reactant basis)
q_i	heat flux based on inside surface area of tube, $HL^{-2}t^{-1}$
q_{iso}	isoteric heat of adsorption, $H(mole)^{-1}$
q_O	heat flux based on outside surface area of tube, $HL^{-2}t^{-1}$
q_p	heat generation potential, dimensionless (see p. 285[1])
q_r	heat removed, Ht^{-1} (prime indicates per mole of reactant basis)
q_v	heat generated per unit volume of tube, $HL^{-3}t^{-1}$
q_{vw}	heat generated per unit volume of tube determined at wall temperature, $HL^{-3}t^{-1}$
q_z	heat flux per unit length, $HL^{-1}t^{-1}$
R, S, ...	products
R	gas-law constant in PVT units, 82.06 $(cm^3)(atm)/$ (g mole)(°K), 0.08204 (liter)(atm)/(g mole)(°K), 10.731 $(ft)(lb_f)/(in)^2$(lb mole)(°R), or 0.7302 (ft^3) (atm)/(lb mole)(°R)
R'	gas-law constant in thermal units, 1.987 g cal/(g mole)(°K) or BTU/(lb mole)(°R)
R_1, R_2	reactivity ratios for copolymerization
R_B	radius of a bend, L
R_b	radius of bed, L
R_d	diffuser throat radius, L
R_{eff}	effective radius of agitation, L
R_j	rate of production of j in mass units, $ML^{-3}t^{-1}$
R_p	radius of spherical particle, L
R_q	$(dq_g/dT)/(dq_r/dT)$
$\mathcal{R}_h$	fouling factor
$\mathcal{R}_p$	resistance coefficient for pipe (see p. 421[1])
r	radial distance from center of bed, L
$r_1, r_2, ...$	rates of reaction No. 1, No. 2, ...
$(-r_A)$	rate of disappearance of reactant A, $(moles)L^{-3}t^{-1}$
$(-\hat{r}_A)$	rate of disappearance of reactant A on a unit mass of catalyst basis, $(moles)M^{-1}t^{-1}$

$(-r_A)_a, (-r_B)_b$	rate of indicated component disappearance per unit volume of indicated phase
r_a	rate of adsorption, $(\text{moles})M^{-1}t^{-1}$
$(\Delta r_A)_P, (\Delta r_A)_T$	rate error caused by pressure and temperature uncertainty, respectively
$(-\hat{r}_A)_s$	rate of disappearance of A based on conditions at catalyst exterior surface, $(\text{moles})M^{-1}t^{-1}$
$(-r_B)_{so}$	mass of solid B consumed per unit time, Mt^{-1}
r_{ck}	rate of coke formation, $(\text{moles})L^{-3}t^{-1}$
r_d	rate of desorption, $(\text{moles})M^{-1}t^{-1}$
r_g	reaction rate in gas phase, $(\text{moles})L^{-3}t^{-1}$
$\mathbf{r}_I$	radial distance from center of particle (or drop), L
(r_I)	rate of generation of new radicals (rate of initiation) $(\text{moles})L^{-3}t^{-1}$
r_L	rate of reaction in liquid phase
$\mathbf{r}_p$	pore radius, L
r_{pr}	rate of polymerization (propagation), $(\text{moles})L^{-3}t^{-1}$
r_R, r_S, r_j	rate of formation of indicated product, $(\text{moles}) L^{-3}t^{-1}$
r_{To}	combined rate in liquid and gas phase
r_t	rate of termination
ΔS°	standard entropy of reaction, $H(\text{mole})^{-1}T^{-1}$
$\Delta S^{\ddagger}$	entropy of activation, $H(\text{mole})^{-1}T^{-1}$; subscript "L" refers to standard state of pressure and temperature of the system, and subscript "p" refers to standard state of 1 atm
ΔS_a	standard entropy of adsorption, $H(\text{mole})T^{-1}$
S_c	coil spacing, L
S_g	total surface area of a porous solid per unit mass, L^2M^{-1}
S_{ℓ}	slope
S_n	stoichiometric ratio of carbon burned per molecule of O_2 consumed
S_R, S_S	selectivity to desired product R and undesired product S (prime designates instantaneous value)
(SV)	space velocity, t^{-1}, defined as volumes or mass
s	number of equidistant centers surrounding each active site
T	temperature of reacting fluid, $^{\circ}K$ or $^{\circ}R$
T_b	bulk fluid temperature
T_{cr}	critical temperature
ΔT_{ck}	temperature drop across coke

$(\Delta T)_D$	temperature increase to double reaction rate
T_e	effluent or exit temperature
T_{eq}	equilibrium temperature
T_F	feed temperature
ΔT_f	temperature drop across film
$\boldsymbol{T_h}$	throw, distance traveled from a diffuser to a point corresponding to a predetermined terminal velocity, feet
T_I	interior catalyst temperature
T_{iso}	isokinetic temperature
T_j	jacket temperature (temperature of cooling or heating medium)
ΔT_M	temperature difference between middle of reactor and wall
T_m	mixture or mean temperature as indicated
T_{op}	optimum temperature for maximum rate
T_0	inlet or initial temperature or base temperature
T_Q	quench temperature
T_r	reduced temperature
T_{R_b}	temperature in packed bed at $\mathbf{r} = R_b$; superscripts f and s refer to fluid and solid, respectively
T_{reg}, T_{re}	temperatures in regenerator and reactor; respectively
T_s	exterior surface temperature of catalyst
T_w	tube wall temperature
ΔT_w	temperature drop across metal wall
T_x	temperature of coaxial gas (see Table 10.9)
t	time, t
t_b	batch mixing time, t
t_D	diffusion time, t
t_f	mixing time with flow but no agitation, t
t_{mix}	mixing time for CSTR, t
t_R	reaction time, t
t_s	time on stream, t
$t_{\Delta Z}$	contact time for length ΔZ, tL^{-1}
U	overall heat-transfer coefficient with driving force of difference between bulk mean reactor temperature and jacket fluid temperature, $H/t^{-1}L^{-2}T^{-1}$
$U_A, U_B, U_C, U_j, \dots$	internal energy of indicated component per mole, $H(\text{mole})^{-1}$
ΔU_A	internal energy of reaction per mole of A reacted, $H(\text{mole})^{-1}$

$U_{f_A}{}^\circ, U_{f_B}{}^\circ, U_{f_C}{}^\circ, \ldots$	standard internal energy of formation at 25°C, $H(\text{mole})^{-1}$
U_v	overall volumetric coefficient of heat transfer, $HL^{-3}t^{-1}T^{-1}$
U_z	point value of overall heat-transfer coefficient, $Ht^{-1}L^{-2}T^{-1}$
U_θ	point value in time of overall coefficient of heat transfer
u	velocity, Lt^{-1}
u'	shear rate, t^{-1}
u_1, u_2	upstream and downstream velocity, respectively, Lt^{-1}
u_B	bubble velocity, Lt^{-1}
u_b	burning velocity, Lt^{-1}
u_{ch}	superficial velocity of fluid at choking, Lt^{-1}
u_f	free settling velocity, Lt^{-1}
u_g	average linear velocity of gas phase in a fluidized bed or gas–liquid system, Lt^{-1}
u_i	interstitial velocity in a bed of catalyst, Lt^{-1}
u_j	jet velocity, Lt^{-1} (see Table 10.9)
u_m	critical minimum superficial velocity to maintain all distributor openings operative, Lt^{-1}
u_{mf}	superficial velocity at point of minimum fluidization velocity, Lt^{-1}
u_o	velocity through orifice or sparger hole velocity, Lt^{-1}
u_p	average linear velocity of particles, Lt^{-1}
u_s	superficial fluid velocity, Lt^{-1}
$(u_s)_g$	superficial gas velocity, Lt^{-1}
u_{sg}	slug velocity, Lt^{-1}
$(u_s)_L$	superficial liquid velocity, Lt^{-1}
u_{sl}	slip velocity, Lt^{-1}
u_{st}	saltation velocity, Lt^{-1}
$\overline{u_{st}}$	saltation velocity for mixed sizes, Lt^{-1}
u_t	terminal velocity, Lt^{-1}
u_{vn}	velocity of vapor based on net area of tower (cross-sectional area less downcomer cross section), Lt^{-1}
u_x	coaxial velocity, Lt^{-1} (see Table 10.9)
u_z	point velocity, Lt^{-1}
V	volume of reactor, L^3
V°	volume of gas adsorbed corrected to 0°C and 760 mm, L^3

V_a, V_b	fractional volumes of indicated phases, volume of phase/total volume of mixture
V_B	volume of bubble, L^3
V_i	volume of reaction mixture below slurry-liquid interface, L^3
V_k	volume of particle, L^3
V_L	liquid volume, L^3
V_m	volume of gas adsorbed as a monolayer corrected to $0°C$ and 760 mm, L^3
V_p	pore volume per unit mass of particle, $L^3 M^{-1}$
V_R	volume of reaction mix, L^3
V_{sg}	volume of slug, L^3
V_T	total aerated or expanded volume, L^3
v	specific molal volume, $L^3(\text{mole})^{-1}$
$v^{\ddagger}$	molal volume of activated state $L^3(\text{mole})^{-1}$
$\Delta v^{\ddagger}$	difference in molal volume between the activated state and the reactants, $L^3(\text{mole})^{-1}$
v_m	mean molal volume, $L^3(\text{mole})^{-1}$
W	mass of catalyst, M
W_a	actual work, $H(\text{mole})^{-1}$
W_d	width of flat diffuser, L
W_f	mass of fluid flowing per unit time, Mt^{-1}
W_m	weight fraction of a given mass
$(-W_s)$	isentropic work, $H(\text{mole})^{-1}$
w	extent-of-reaction factor (see p. 174[1])
w_2	weight fraction of polymer in polymer-rich phase
w_I	impeller blade width, L
(w_I)	weight fraction initiator in monomer
w_{pr}	weight fraction of polymer in solution
w_x	weight fraction of size fraction x
X_A, X_R, $\ldots$	moles of indicated component converted per mole of A charged or fed
$X_A{}^*$	equilibrium conversion of A
X_{A_T}, X_{R_T}, $\ldots$	moles of indicated component converted per mole of total feed or charge
X_e	conversion at reactor discharge
X_1	conversion level at reactor inlet
X_0	conversion level in reactor feed or charge
x_A, x_B, $\ldots$	mole fraction of indicated component in liquid phase
x_n	mole fraction in liquid from tray n

Y_C	fractional carbon remaining on catalyst, moles carbon remaining per mole carbon initially present
Y_G, Y_L	ratio of ΔP gradient for mixed phase to that for gas alone (G) or liquid alone (L) (see Fig. 14.8)
Y_R	yield of desired product R (prime designates instantaneous value)
$y_A, y_B, \ldots, y_R, y_S$	mole fraction of indicated component in gaseous phase (subscript i = interfacial value, subscript e refers to effluent, subscript 0 refers to feed or inlet value)
y^+	mole fraction of a component in vapor in physical equilibrium with that component in liquid
y_b	mole fraction in rich gas entering bottom of tower
y_f	mole fraction factor for mass transfer, see Eq. 11.15, dimensionless
y_n	mole fraction of a vapor component from tray n.
y_t	mole fraction in lean gas leaving top of tower
Z	longitudinal distance along a reactor, L
Z''	length of poisoned zone in catalyst, L
Z_c	height of coil from tank bottom
Z_F	freeboard height, L
Z_I	height of impeller from tank bottom, L
Z_L	liquid height in a stirred or sparged vessel, L
Z_{mf}	height of bed at point of minimum fluidization velocity, L
Z_s	height of suspended solids, L
Z_{sg}	height of equivalent cylindrical slug having same volume as actual slug (see Table 13.8)
Z_T	distance between set of nozzles or height of chamber, ft (see Table 10.9)
Z_v	vertical distance between pressure taps, L
z	distance along film
z_m	compressibility factor for mixture

Greek Letters

α, β	various exponents and constants defined at point of use
α_a	projected area of molecule adsorbed on surface, L^2 (molecule)$^{-1}$
α_B	dimensionless statistical constant in Ergun equation

α_{ck}	fraction of coke generated that deposits on wall
α_p	fraction of catalyst surface poisoned, coked, or deactivated in some manner
β	solids angle of repose
β_B	constant in Ergun equation
β_d	constant in deactivation equation
Γ_A	ratio of mass-transfer rate of A to total rate of transfer of all components
γ'	constant in deactivation equation
$\gamma_A, \gamma_B, \ldots, \gamma_R, \gamma_S, \ldots, \gamma^{\ddagger}$	activity coefficients for indicated components
$\gamma_A{}^*, \gamma_B{}^*$	activity coefficient for indicated components that includes v_m
δ	fractional loss of selectivity
δ_A	change in moles per mole of A reacted
δ_c	thickness of combustion wave, L
δ_f	film thickness, L
δ_p	fraction of total pore length occupied by a plug of liquid
δ_v	fractional change in volume upon polymerization
δ_X	experimental error in conversion
ε	void fraction of a bed of catalyst
ε_b	fraction of fluidized bed occupied by bubbles
ε_{ch}	void fraction at choke
ε_l	liquid fraction of system volume below interface
ε_{mf}	void fraction at minimum fluidization velocity
ε_p	particle void fraction
ε_r	roughness of pipe
ε_w	screen porosity or void fraction
ζ_j	stoichiometric coefficient for species j relative to A, moles of j/mole of A in stoichiometric relation
η	effectiveness factor
η_I	inherent viscosity, $\ln (\eta_r) g^{-1} (100 \text{ ml})$
η_O	effectiveness factor including bulk transport resistances
η_r	relative viscosity, polymer solution viscosity/solvent viscosity
$[\eta]_T$	intrinsic viscosity, dl./g
θ	angle

θ_a	fraction of surface covered at equilibrium adsorption
θ_s	fraction of sites available at any time
κ	transmission coefficient (see p. 17[1])
κ°	heat capacity ratio, c_p/c_v
κ_f	flow parameter (see Fig. 14.10 and Eq. 14.29)
Λ_t	fraction of fluid remaining longer than time t
λ	thermal conductivity, $HL^{-1}t^{-1}T^{-1}$
λ_a	axial thermal conductivity of catalyst bed, $HL^{-1}t^{-1}T^{-1}$
λ_{ck}	thermal conductivity of coke, $HL^{-1}t^{-1}T^{-1}$
$\lambda_e, \lambda_{ea}, \lambda_{ew}$	effective thermal conductivity general, of an aggregate of particles, and of an aggregate next to wall, $HL^{-1}t^{-1}T^{-1}$
λ_f	thermal conductivity of fluid, $HL^{-1}t^{-1}T^{-1}$
λ_g	gas conductivity, $HL^{-1}t^{-1}T^{-1}$
λ_I	effective thermal conductivity of catalyst particle, $HL^{-1}t^{-1}T^{-1}$
λ_r	effective radial thermal conductivity in operating catalyst bed, $HL^{-1}t^{-1}T^{-1}$
λ_r^f	fluid contribution to effective radial thermal conductivity, $HL^{-1}t^{-1}T^{-1}$
λ_r^s	solid contribution to effective radial thermal conductivity, $HL^{-1}t^{-1}T^{-1}$
λ_w	thermal conductivity of wall, $HL^{-1}t^{-1}T^{-1}$
μ	viscosity, $ML^{-1}t^{-1}$
μ_b	viscosity of bulk fluid, $ML^{-1}t^{-1}$
μ_{cr}	critical viscosity, $ML^{-1}t^{-1}$
μ_f	fluid viscosity, $ML^{-1}t^{-1}$
μ_g	viscosity of gas, $ML^{-1}t^{-1}$
μ_L	viscosity of liquid, $ML^{-1}t^{-1}$
μ_m	viscosity of mixture, $ML^{-1}t^{-1}$
μ_{mi}	viscosity, micropoise
μ_r	reduced viscosity, μ/μ_c
μ_w	viscosity at wall, $ML^{-1}t^{-1}$
ν	kinematic viscosity, L^2t^{-1}
$\nu_A, \nu_B, \nu_j, \ldots, \nu_R, \nu_S, \ldots, \nu^{\ddagger}$	fugacity coefficients for indicated components
ξ	instantaneous number average degree of polymerization
ρ_b	bulk density, ML^{-3}
ρ_{ck}	density of coke, ML^{-3}
ρ_F	molar density of feed, (moles)L^{-3}

ρ_f	fluid density, ML^{-3}
ρ_g	gas or vapor density, ML^{-3}
ρ_L	density of liquid, ML^{-3}
ρ_p	particle density, ML^{-3}
ρ_{pr}	polymer density, ML^{-3}
ρ_s	solid density (excludes pore volume), ML^{-3}
σ	surface tension, Mt^{-2} or dynes/cm
σ_{AB}	distance between two molecules at collision when considered as rigid spheres, L
σ_c	critical surface tension, Mt^{-2}
σ_e	$\sigma_{AB}\mathbf{p}^{\frac{1}{2}}$, related to collision cross section, L
τ	mean residence time for constant density flow, equivalent to holding time (reactor volume divided by volumetric flow rate) in plug flow and CSTR (backmixed reactor), t
Υ	tortuosity factor
Φ_s	effectiveness factor modulus in terms of observables (see Eq. 3.23)
ϕ_d	drop diffusion modulus (see Eq. 15.3)
ϕ_f, ϕ_q	correction constants in rate equations for gaseous and liquid reaction systems, respectively (see p. 37[1])
ϕ_L	Thiele modulus for flat slab (see p. 134[1])
ϕ_s	Thiele effectiveness factor modulus for spherical particle (see Eq. 3.14, 121[1])
ϕ_{sp}	sphericity factor (see Eqs. 11.17 and 13.2)
ψ	instantaneous activity, actual rate/rate at zero time or at start up
ω_A	$\delta_A y_{A_0}$
ω_e	(equivalent length of pipe)/(actual length)
ω_i	dispersed phase mixing frequency or interaction rate, volume mixed/(time)(volume of dispersed phase)

Frequently used subscripts:

e	effluent or product condition
f	fluid
i	value at interface; also any free radical
j	any component; also used to refer to jacket
LM	logarithmic mean
0	initial or feed condition
s	refers to surface of particle unless noted otherwise

Frequently used superscripts:

f   fluid
°   standard state unless otherwise defined
•   final discharge or product value
s   refers to surface of particle unless noted otherwise
‡   activated state
*   value at chemical equilibrium unless noted otherwise

INDEX